A BIOGRAPHICAL INDEX
OF
BRITISH ENGINEERS
IN THE
19th CENTURY

Garland Reference Library of Social Science (Vol. 5)

A BIOGRAPHICAL INDEX
OF
BRITISH ENGINEERS
IN THE
19th CENTURY

compiled by
S. P. Bell

Garland Publishing, Inc., New York & London

1975

Library of Congress Cataloging in Publication Data

Bell, S   Peter.
   A biographical index of British engineers in the
19th century.

   (Garland reference library of social science ; 5)
   1.   Engineers--Great Britain--Biography.
I.   Title.
TA157.B44   016.62'00092'2   75-5114
ISBN 0-8240-1078-7

CONTENTS

PREFACE

A study of the British engineering profession as it emerged
in the years following the development of the railways in
the 1830's is crucial to the understanding of both economic
history and the history of technology. And yet the vast majority of the men responsible for the engineering achievements
of the Industrial Revolution remain unknown figures, their
only memorial being very often obituary notices found in
the contemporary engineering press. This index has therefore
been compiled to give access to the obituary notices of some
3,500 British engineers who died on or before the year 1900.
The date 1900 has been chosen because it is thought to include within it the deaths of the vast majority of those persons
responsible for the major developments in British engineering-
particularly in the second quarter of the 19th century- and for
its spread and consolidation throughout the world in the second
half of the 19th century.

Over forty contemporary engineering journals were scanned for
possible inclusion in this index, of which only twenty were
found to contain regular obituary notices during this period.
These included the journals of most of the major engineering
institutions, as well as the general journals Engineer and
Engineering.
Because of the size of the undertaking, it was not practicable
to follow through further references thrown up by the obituaries themselves. For example, references are often made to
the Times newspaper, as well as to the journals of such non-
engineering societies as the Royal Society of Edinburgh, the
Royal Society of Arts, the Royal Geographical Society, &c..
The only two works to which reference has been made (usually to
supplement dates)are the invaluable general biographical dictionary compiled by Frederick Boase, Modern British Biography
(7 vols, 1892-1921: reprinted London, 1965); and to the
Dictionary of National Biography(22 vols, 1885-1900). Where an
engineer is noted in Boase, the letter (B) follows his occup-
ation in this index, and where noted in DNB the letter (D).

While the reader must draw his own conclusions as to why the
deaths of certain important engineers were omitted by their
contemporaries, it should be noted here that great difficulty
was experienced with the problem of non-engineers whose obituaries did appear in the engineering press. These included not
only the obvious categories(such as painters, doctors or
musicians) and honorary members (royalty, politicians and some
peers), but also many pure scientists (physicists, mathematic-
ians and astronomers) as well as geologists and chemists. It
was finally decided to include here not only all branches of engineering proper as these emerged as distinct entities during
the course of the century, but also those persons whose activ-
ities obviously connected them closely with engineering _

such as inventors, some entrepreneurs and company directors, architects, patent agents, &c. Non-British engineers have, by virtue of their number, been excluded throughout, as have attempts to locate British obituaries in non-British journals.

While the actual obituaries themselves vary greatly in size and quality, the only distinction drawn here has been to identify with an asterisk those which contain only a passing reference to the death of an individual. Illustrated obituaries are also noted without distinction as to whether this is a photograph or wood engraving.
An attempt has also been made to identify the specialist field of each engineer, but because of the astonishingly varied and diverse careers of many of them, it is likely that the compiler's attempts to do this satisfactorily may be queried. If so, he will be glad to acknowledge any such corrections, together with any other omissions or mistakes, with a view to incorporating them into any future revision.

Finally, I should like to acknowledge here my indebtedness to Professor Eugene S. Ferguson, whose Bibliography of the History of Technology (MIT Press, 1968) gave me the idea for this index in the first place. Since then, Professor Ferguson has given me his active encouragement and support in completing the project - albeit at a distance of 3,000-odd miles. My colleague Miss Dilys Bateman also helped me by reading the proofs, both at manuscript and typescript stage; while in the last analysis, it proved to be my wife who pushed the whole through to its completion - in spite of an almost total disinterest in the subject matter. Without them, a work which I believe is long overdue would have been impossible.

University of Salford,   Samuel Peter Bell
Salford M5 4WT.
England.

CEAJ Civil Engineers and Architects Journal.
   v.1 1837 - v.31 1868.

Elec Electrician.
   v.1 1878-

Elec Rev Electrical Review.
   v.30 1892- . Formerly Telegraphic Journal.

Eng Engineer.
   v.1 1866-

Engng Engineering.
   v.1 1870-

Iron Iron.
   v.1 1873 - v.41 1893. Formerly Mechanics
   Magazine.

JIEE Journal of the Institution of Electrical Engineers.
   v.18 1889- Formerly Journal of the Society of
   Telegraph Engineers.

JISI Journal of the Iron and Steel Institute.
   v.1 1869-

JSTE Journal of the Society of Telegraph Engineers.
   v.1 1872 - v.17 1888. Continued as Journal of the
   Institution of Electrical Engineers.

Mech Mag Mechanics Magazine.
   v.1 1823 - v.92 1872. Continued as Iron.

NECIES North-East Coast Institution of Engineers and Ship-
   builders - Transactions.
   v.1 1884/5-

PAMSES Proceedings of the Association of Municipal and Sanitary Engineers and Surveyors.
   v.1 1873/4 - v.16 1890. Continued as Proceedings
   of the Incorporated Association of Municipal and
   County Engineers.

PICE Proceedings of the Institution of Civil Engineers.
   v.1 1837-

PIMCE Proceedings of the Incorporated Association of Municipal and County Engineers.
   v. 17 1890- . Formerly Proceedings of the Association of Municipal and Sanitary Engineers and Surveyors.

PIME   Proceedings of the Institution of Mechanical
            Engineers.
              v.1 1847-

Prac Mag   Practical Magazine.
              v.1 1873 - v.7 1877.

TCMCIE   Transactions of the Chesterfield and Midland
            Counties Institute of Engineers.
              v.1 1871/3 - v.17 1888/9.

TFIME   see TIME

TIESS   Transactions of the Institution of Engineers and
            Shipbuilders of Scotland.
              v.1 1857/8-

TIME   Transactions of the (Federated)Institution of
            Mining Engineers.
              v.1 1889-

TINA   Transactions of the Institution of Naval Architects.
              v.1 1860-

TMAE   Transactions of the Manchester Association of Eng-
            ineers.
              v.1 1891-

TNEIMME   Transactions of the North of England Institute of
            Mining and Mechanical Engineers.
              v.1 1852/3 - v.38 1889.

TSE   Transactions of the Society of Engineers.
              v.1   1857 -

Abbey, John Henry, 1831-80, surveyor.
    PICE - 63, 301.

Abbott, John Hallen, 1831-84, railway engineer in India.
    PICE - 79, 360-361.

Abbott, Samuel, 1842-90, railway engineer.
    PICE - 102, 329-330.

Abernethy, George, 1815-65, railway engineer.
    PICE - 25, 522-523.

Abernethy, Harold William, 1858-95, Manchester Ship Canal engineer(B).
    PICE - 122, 363-366.

Abernethy, James, 1814-96, civil engineer.
    Eng - 81, 271.
    Engng - 61, 360*, 384-385.
    PICE - 124, 402-407.
    PIME - 1896, 90-91.

Abernethy, James, Jun., 1844-97, design engineer.
    PICE -130, 312.

Acton, William Richard,   -1899, municipal engineer.
    PICE - 137, 439.

Adam, William, 1837-96, locomotive engineer.
    Eng - 81, 402+port.
    Engng - 61, 472-473.
    TIESS - 40, 249.

Adams, Edward, 1814-75, industrial architect.
    PICE - 41, 221-222.

Adams, George, 1829-91, manufacturing engineer.
    JISI - 39, 247-248.

Adams, George Frederick, 1842-84, ironworks manufacturer.
    Engng - 38, 386.
    JISI - 27, 537-538.
    PICE - 81, 320-321.

Adams, John,   -1850, ---
    PICE - 10, 95*.

Adams, John Henry,   -1877, ---
    Eng -44, 63*.

Adams, Thomas, 1826-82, inventor & manufacturer(B).
    PIME - 1883, 1.

Adams, W   ,   -1894, locomotive superintendent.
    Eng - 77, 571+port.

Adams, William, 1813-86, civil & mining engineer.
    PICE - 87, 416-417.

Adams, William Alexander, 1821-96, railway carriage manufr.
    PICE - 124, 436-437.
    PIME - 1896, 91-92.

Adams, William Bridges, 1797-1872, inventor & carriage
    builder(B, D).
    Eng - 34, 52.
    Engng - 14, 63-64.

Adamson, Daniel, 1820-90, boilermaker(B).
    Eng - 69, 56.
    Engng - 49, 66-68.
    Iron - 35, 57.
    JISI - 36, 168-175. Photo. in General Index, 1902.
    PICE - 100, 374-378.
    PIME -   167-171.

Adamson, James Grey, 1839-81, engineer in Chile.
    PICE - 67, 403-405.

Adcock, Charles, 1860-98, municipal engineer.
    PICE - 135, 362-363.

Addenbrooke, Edwin, 1845-87, gas engineer.
    PICE - 89, 490.

Addis, William John, 1832-94, engineer in India & inventor.
    TSE - 1897, 201.

Addison, Capt. John Copley, 1852-87, Royal Engineer.
    PICE - 91, 455-457.

Addy, John, 1847-96, drainage engineer.
    PICE - 127, 361-362.

Adie, Patrick, 1827-86, instrument maker.
    PICE - 86, 367-368.

2

Adley, Charles Coles, 1828-96, civil engr & publisher(B).
    PICE - 125, 414-416.

Agnew, George Alexander, 1853-99, shipbuilder.
    Eng - 88, 150.
    TIESS - 42, 402.

Aher, David, 1780-1843, civil engineer.
    PICE - 3, 14.

Ainger, Thomas Edward,   -1857, railway engr in Brazil.
    PICE - 17, 107*.

Ainslie, William George, 1832-93, iron & steel manufr.
    Iron - 41, 143*.
    JISI - 43, 172.

Aird, John,   -1876, contractor.
    Engng - 21, 382.

Aitchison, George, 1792-1861, architect & surveyor(B).
    PICE - 21, 567-571.

Aiton, William, c.1820-93, civil engineer, contractor(B).
    Iron - 41, 176.
    TIESS - 36, 317.

Alexander, Thomas, 1817-1900, engineer in Spain.
    Engng - 69, 13.

Alexander, William, 1818-86, Inspector of Mines(B).
    Engng - 41, 81.
    TIESS - 29, 219.

Alford, Richard Francis, 1849-82, mechanical engineer(B).
    PICE - 74, 291-292.

Allan, Alexander, 1809-91, locomotive engineer.
    Eng - 71, 471.
    PIME - 1891, 289-290.

Allan, Archibald Binnie, 1850-98, municipal engineer.
    Engng - 65, 304.
    TIESS - 41, 378.

Allan, J   Carruthers, 18(66)-91, telegraph engineer.
    Elec - 26, 530*.

Allan, James, 18(11)-74, shipping director(B).
    PICE - 39, 283-285.

Allan, James, 1837-92, steel manufacturer.
    JISI - 42, 304.

Allcard, William, 1809-61, railway engineer.
   PICE - 21, 550-552.

Allen, Alfred Evans, 1846-93, marine surveyor.
   PIME - 1893, 489.

Allen, Harry, 1853-99, steel manufacturer.
   Engng - 67, 288.
   JISI - 55, 261.

Allen, Henry, 18(44)-1900, civil engineer.
   Eng - 90, 417*.

Allen, James, 1824-64, brass & copper manufacturer.
   PIME - 1865, 13.

Allen, John Prosser, 18(26)-99, naval engineer.
   Eng - 88, 546.

Allen, William Daniel, 1824-96, steel manufacturer(B).
   Elec Rev - 39, 572*.
   Eng - 82, 449+port.
   Engng - 62, 562-563+port.
   JISI - 50, 255-257.

Alley, Stephen, 1840-98, engine manufacturer.
   Elec Rev - 42, 450*.
   Eng - 85, 306.
   Engng - 65, 368*, 409.
   JISI - 53, 311-312.
   PIME - 1898, 132-133.

Allibon, George,   -1886, marine superintendent.
   Eng - 62, 323, 347.

Allison, John, 1838-94, Manchester City Engineer(B).
   PICE - 116, 351-352.
   PIMCE - 20, 380-381.

Allport, Sir James Joseph, 1811-92, railway manager(B).
   Eng - 73, 365.
   Engng - 53, 531-532.
   Iron - 39, 386.

Allport, Samuel Blakemore, 1823-99, gun manufacturer(B).
   Eng - 88, 428.

Alsing, Gustav Valentine,   -1896, sanitary engineer.
   Eng - 82, 403.
   PICE - 127, 387.

Alty, Henry, 1844-82, municipal engineer.
   PICE - 71, 420-421.
   PAMCES - 8, 228.

Amos, Charles Edwards, 1805-82, inventor(B).
    Eng - 54, 126.
    PICE - 71, 387-395.
    PIME - 1883, 1-7.

Amos, James Chapman, 1836-1900, consulting engineer.
    PIME - 1900-1, 461.

Anderson, Alexander, 1853-96, dock engineer.
    PICE - 127, 388.

Anderson, Henry John Card, 1835-91, contractor overseas.
    PICE - 106, 318-319.
    PIME - 1891, 472-473.

Anderson, Herbert Goulburn, 1842-68, railway inspector.
    PICE - 28, 618.

Anderson, James, 1811-97, shipping manager(B).
    Eng - 84, 260.

Anderson, Sir James, 1824-93, telephone company director(B).
    Elec - 31, 30,38,85.
    Elec Rev - 32, 564, 594.
    Eng - 75, 412.
    Engng - 55, 675.
    Iron - 41, 406.

Anderson, James,   -1899, steel works manager.
    Eng - 88, 484.

Anderson, James Arthur, 1843-99, railway engr in India.
    PICE - 139, 349-350.

Anderson, Sir John, 1814-86, armaments manufacturer(B).
    Eng - 62, 95.
    Engng - 42, 119.
    Iron - 28, 111.
    PICE - 86, 346-353.
    PIME - 1886, 460-461.

Anderson, John Philip Cortlandt,1831-93, Public Works engr in
    India.
    PICE - 115, 381-382.

Anderson, Mathew,   -1883, marine superintendent.
    Engng - 36, 152.

Anderson, Samuel, 1847-90, ironworks manager.
    PIME - 1890, 554.

Anderson, W   D   ,   -1842, ---
    PICE - 2, 13.

Anderson, Sir William, F.R.S., 1835-98, Director-General
    of Ordnance(B,D).
    Eng - 86, 587+port.
    Engng - 66, 787-789+port.
    JIEE - 28, 665.
    JISI - 54, 324-325.
    PICE - 135, 320-326.
    PIME - 1898, 696-701.
    TSE - 1898, 240.

Andrew, J   Arnold,   -1891, steel manufacturer.
    Eng - 71, 380.

Andrew, John Henry, 1824-84, steel works manager.
    Iron - 24, 251.
    JISI - 25, 554-555.

Andrew, Peter, 1837-86, locomotive engineer.
    Engng - 41, 323*.

Andrew, Sir William Patrick, 1806-87, railway & telegraph
    engineer in India(B).
    Elec Rev - 20, 260*.
    Eng - 63, 218.
    Iron - 29, 232.

Andrews, Charles, 1829-93, naval engineer(B).
    Eng - 75, 246.
    PICE - 113, 327-328.

Andrews, George Robert,   -1899, municipal engineer.
    PIMCE - 25, 479*.

Andrews, James, 1827-97, contracting engr in America.
    PICE - 131, 390-391.

Andrews, John Orme, 1832-1900, naval engineer.
    PICE - 141, 333-334.

Angus, Robert Nicoll, 1811-75, locomotive superintendent.
    PIME - 1876, 17.

Anley, George Augustus d'Auvergne, 1833-1900, Public Works
    engineer in India(B).
    PICE -   141, 334-335.

Annand, James,   -1889, railway engineer in Japan.
    Iron - 33, 338.

Anningson, William Philipson, 1859-99, dock engineer.
    PICE - 139, 372-373.

Ansell, George Frederick, 1826-81, mining chemist.
    Iron - 17, 14.

Anstead, David Thomas, F.R.S.,1814-80, Professor of Geology &
    consulting mining engineer(B, D).
    Eng - 49, 393.
    Iron - 15, 410.

Anstice, William Reynolds, 1807-81, ironmaster(B).
    JISI - 19, 577*.

Appleby, Henry, 18(37)-89, consulting railway engineer.
    Eng - 68, 513.

Applegath, Augustus, 1787-1871, inventor & printer(B).
    Engng - 11, 119-120.

Appold, John George, F.R.S., 1800-65, inventor of centrifugal
    pump(B,D).
    Eng - 20, 147.
    PICE - 25, 523-525.

Arch, Arthur Joseph Edwin, 1862-91, surveyor overseas.
    PICE - 104, 307-308.

Archbold, John, 1838-97, colliery engineer.
    PIME - 1897, 513.

Archbold, William, 1868-96, draughtsman.
    NECIES - 13, 269.

Archbould, Ralph, 1856-89, electrical works manager.
    PICE - 101, 307.

Archdeacon, Staff-Capt.William Edwin, R.N., 1839-93, marine
    surveyor(B).
    PICE - 113, 360-361.

Armitage, William James, 1819-95, entrepreneur(B)
    PIME - 1895, 308.

Armstrong, Capt.   ,   -1897, railway dockmaster.
    Eng - 83, 605*.

Armstrong, Col.   ,   -1894, electrical engineer.
    Elec - 34, 125*.

Armstrong, George Frederick, 1842-1900, Professor of Engineering,
    Edinburgh University(B).
    Elec Rev - 47, 836*.
    Eng - 90, 523.
    Engng - 70, 674.
    JISI - 58, 387-388.
    PICE - 144, 308-312.
    PIME - 1900, 621-622.

Armstrong, John, 1775-1854, municipal engineer of Bristol(B).
    PICE - 14, 147-148.

Armstrong, Joseph, 1816-77, locomotive superintendent(B).
    Eng - 43, 400*, 409.
    Engng - 23, 447.
    PICE - 49, 255-258.
    PIME - 1878, 9-10.

Armstrong, Joseph, Jun., 1856-88, railway engineer.
    PIME - 1888, 153.

Armstrong, Robert,   -1868, mechanical engineer.
    Eng - 26, 5,33,82.

Armstrong, Thomas William de Butts, 1826-77, engr in India(B).
    PICE - 51, 261-265.

Armstrong, William, 1812-96, mining engineer(B).
    TFIME - 14, 171-172.

Armstrong, Sir William George, F.R.S., Baron Armstrong, 1810-
    1900, armaments manufacturer(D).
    Elec Rev - 48, 26.
    Eng - 90, 630.
    Engng - 70, 833*.
    JISI - 59, 313-315.
    PICE - 147, 417-412.
    PIME - 1900, 696-701.
    Prac Mag - 4, 81-88+port.
    TIME - 21, 176-188+port. By H.Palin Gurney.
    TINA - 43, 353-354.
    TSE - 1900, 269-270.

Armstrong, William Henry, 1865-95, water engr in Calcutta.
    PIME - 1895, 532.

Arntz, Robert Richard, 1815-82, builder & surveyor(B).
    PICE - 70, 431-432.

Arthur, John Frederick, R.N., 18(71)-1900, naval engineer.
    Eng - 90, 393*.

Ashbury, John, 1806-66, wagon manufacturer(B).
    PIME - 1867, 14.

Ashby,   ,   -1870, ---
    Eng - 29, 202*.

Ashcroft, Peter, 1809-70, railway engineer.
    Eng - 29, 140.

Ashmead, Frederick, 1825-98, sanitary engineer.
    PICE - 135, 326-327.
    PIMCE - 25, 477-478. Port. frontis. vol 18.

Ashwell, Frank, 1855-96, manufacturing engineer.
    PIME - 1896, 596.

Ashwell, James, 1799-1881, railway engineer(B).
Eng - 52, 31.
Engng - 32, 46.
PICE - 66, 372-375.

Asquith, Edmund, 18(59)-92, hydraulic engineer.
TMAE - 2, 277.

Atchison, Arthur Turnour, 1848-91, consulting engineer.
PICE - 105, 320-322.

Atherton, Charles, 1805-75, consulting engineer(B).
PICE - 42, 252-255.

Atkinson, Charles Robert, 1826-89, railway engineer.
PICE - 100, 378-379.

Atkinson, John Staines, 1822-64, mechanical engineer.
PICE - 25, 525-526.

Atkinson, T   ,   -1845, ---
PICE - 5, 5*.

Auld, David, 1809-99, inventor.
TIESS - 42, 402-403.

Austen, Stanley, 18(25)-1900, shipbuilder.
Eng - 90, 304.

Austin, Charles Edward, 1819-93, railway engr overseas(B).
PICE - 113, 329-331.

Austin, Henry, 1809-61, civil engineer(B).
CEAJ - 24, 346.
Eng - 12, 249*.

Austin, R   ,   -1891, Secretary, Amalgamated Society of
Engineers.
Engng - 52, 368-369.

Aveling, Thomas, 1824-82, steam roller manufacturer(B).
Eng - 53, 180.
Engng - 33, 228-229.
Iron - 19, 206-207.
JISI - 21, 646-649.
PICE - 73, 350-355.
PIME - 1883, 7-11.

Avern, Frederick Morris, c.1840-86, Public Works engineer in
India.
PICE - 85, 393-394.

Avery, Thomas, 1813-94, weighing machine manufacturer(B).
Eng - 77, 153.

Awdry, Maj. Ambrose, 1844-85, Royal Engineer.
    PICE - 81, 340-342.

Aylmer, John, 18(41)-97, telegraph engineer in France.
    Elec Rev - 40, 116.

Ayrton, Capt. Frederick, 1812-73, civil engineer(B).
    PICE - 38, 306-309.

Aytoun, Robert, 1799-1876, amateur engineer(B).
    PICE - 47, 299-300.

Badderley, Lt.-Col. John Fraser Loddington, 1826-62, Royal
    Artillery.
    Eng - 23, 283.
    PICE - 22, 634-636.

Bage, W   ,   -1887, shipping director.
    Engng - 43, 44*.

Bagnall, Charles, 1827-84, ironmaster(B).
    JISI -   25, 84.

Bagnall, James, 1804-72, coal and ironmaster.
    PICE - 36, 281-282.

Bagnall, William, 1797-1863, coal and ironmaster.
    PICE -24, 540-541.
    PIME - 1864, 13.

Bagot, Alan Charles, 1856-85, mining engineer(B).
    PICE - 81, 337-339.
    PIME - 1885, 300-301.

Bagshawe, John James, 1835-75, steel manufacturer.
    PIME - 1876, 17*.

Bailey, Crawshay, 1789-1872, ironmaster(B).
    Eng - 32, 25, 37*.

Bailey, Crawshay, Jun., 1821-87, ironmaster.
    Engng - 43, 373*.

Bailey, John, 1814-84, manufacturer.
    Iron - 24, 99.

Bailey, John, 1839-92, contractor.
    PICE - 111, 364-366.

Bailey, William,   -1882, shipbuilder.
    Engng - 35, 22.

Bailey, William, 1853-95, mechanical engineer.
    PIME - 1895, 532.

Baillie, Robert, 1818-99, marine engr & shipbuilder(B).
    Eng - 87, 351.
    Engng - 67, 449.
    TSE - 1899, 259-260.

Bain, Alexander, 1810-77, telegraph engineer(B,D).
    Eng - 42, 22.
    Engng -23, 54.

Bain, Sir James, 1818-98, coal and ironmaster(B).
    JISI - 53, 312.

Bain, Capt. John,   -1895, Board of Trade official.
    TIESS - 39, 273.

Bainbridge, Emerson Muschamp,   -1892, ironmaster.
    Iron, 39, 185*.

Baird, Francis, 1802-64, foundry owner in Russia.
    PICE - 30, 428-429.

Baird, James, 1802-70, ironmaster.
    Engng - 9, 74-75.

Baird, James, 1802-76, ironmaster(B,D).
    Engng - 21, 534, 554.
    JISI - 8, 278, 511.

Baird, Robert, 1806-56, ironmaster(B).
    Eng - 2, 454*.

Baker, Charles Bernard, 1832-81, consulting engineer.
    PICE - 65, 364.

Baker, James Philip, 1825-83, Inspector of Mines.
    TCMCIE - 12, 17.

Baker, Samuel, 1822-81, manufacturing engineer.
    PIME - 1882, 1.

Baker, W   S   Graff, 18(75)-97, telegraph engineer.
    Elec - 39, 209*.

Baker, William, 1817-78, railway engineer(B).
    Eng - 46, 462.

11

JISI - 12,293-294.
PICE - 55, 315.

Baker, <u>Gen. Sir</u> William Erskine, 1808-81, Royal Engineer(B).
    Iron - 18, 524.

Baldry, James Danford, 1816-1900, consulting engineer.
    Engng - 69, 261-262+port.
    PICE - 143, 309-311.

Baldwin, John, 1824-91, surveyor.
    PICE - 107, 416-418.

Baldwin, Martin, 1788-1872, mechanical engineer(B).
    PIME - 1873, 16-17.

Baldwin, Thomas, 1822-84, consulting engineer.
    PIME - 1885, 71.

Balfour, James Melville, 1831-69, marine engineer(B).
    Eng - 9, 125*.
    PICE - 31, 200-202.

Ball. H   W   ,   -1887, engineering manufacturer.
    TIESS - 31, 236*.

Ball, John,   -1847, zinc mill proprietor.
    PICE - 7, 15*.

Ballantine, James Cunningham Rollo Bowman, 1865-92, foundry
    chairman.
    TSE - 1892, 257.

Ballard, Stephen, 1804-90, contractor.
    Eng - 70, 476.
    PICE - 104, 288-291.

Baly, Price Richard, 1819-75, amateur engineer.
    PICE - 43, 296-297.

Bamford, Henry, 18(18)-96, agricultural implement manufr.
    Eng - 82, 478.

Bampton, Augustus Hamilton, 1823-57, civil engineer(B).
    PICE - 17, 92-94.

Bancroft, Henry, 1834-1900, consulting engineer.
    PICE - 141, 335-336.

Banfield, Edward, 1837-71, railway engr in South America.
    PICE - 36, 282-284.

Banister, Frederick Dale, 1823-97, railway engineer(B).
    Eng - 84, 650.

Engng - 65, 21.
JISI - 53, 312.
PICE - 131, 359-361.

Banks, Edward Nevill, 1845-96, sanitary engineer.
PICE - 126, 400.

Bannatyne, Neil,   -1892, telegraph company director.
Elec - 28, 421*.
Elec Rev - 30, 263.

Banner, Edward Gregson, 18(13)-90, sanitary engineer.
TSE - 1890, 214.

Bantock, Thomas, 1823-95, railway agent.
JISI - 47, 259-260.

Barbenson, Robert Thomas Olivier, 1845-93, dock engineer.
PICE - 116, 374-375.

Barber, Edmund Scott, 1812-54, civil & mining engineer.
PICE - 14, 126-127.

Barber, Edmund Scott, Jun., 1845-96, civil engr overseas.
PICE - 124, 407-409.

Barber, James, 18(22)-96, municipal surveyor.
PIMCE - 23, 478.

Barber, Thomas,   -1893, mining engr & colliery owner.
TIME - 7, 684.

Barber, William, 1817-76, civil engineer.
PICE - 47, 300-301.

Barbour, Thomas, 18(37)-96, ironworks manager.
JISI - 51, 309.

Barclay, Andrew, 1817-1900, manufacturing engineer.
Eng - 89, 437.
Engng - 69, 547.

Barclay, Arthur, 18(45)-1900, ---
Eng - 89, 541*.

Barker, E   D   , 1844-89, railway engineer.
Engng - 48, 429*.

Barker, Frederick William, 1862-1900, ventilating engineer.
PIME - 1900, 323.

Barker, George J   ,   -1892, ironmaster.
Iron - 39, 405.

13

Barker, S   ,   -1892, ironfounder.
    Iron - 40, 451.

Barker, William, 1817-78, railway & consulting engineer.
    PICE - 55, 315-317.

Barkley, John Trevor, 1826-82, colliery manager(B).
    JISI - 21, 651-653.

Barlow, Peter William, F.R.S., 1800-85, civil engineer(B,D).
    Eng - 59, 422.
    Iron - 25, 495.
    PICE - 81, 321-323.

Barlow-Massicks, Thomas Gibson, 1862-99, civil engr in USA.
    PICE - 136, 357-358.

Barnes, John, 1798-1852, marine engineer(B).
    CEAJ - 15, 399.
    PICE - 12, 140-148.

Barnes, Robert, 1864-94, Public Works engineer in India.
    PICE - 118, 455-456.

Barnes, Thomas Hammerton,   -1892, manufacturing engineer.
    Iron - 39, 163.

Barnett, Edward William,   -1895, contractor.
    Engng - 59, 111.

Barnett, Francis Thomas,   -1896, industrial chemist.
    JISI - 51, 309.

Barningham, James, 18(15)-84, ironmaster.
    Iron - 23, 229.

Barningham, William, 1826-81, ironmaster(B).
    JISI - 21, 657-658.

Barnwell, Richard, 18(49)-98, shipbuilder.
    Eng - 85, 238.
    Engng - 65, 304.

Barrass, Thomas,   -1868, colliery surveyor.
    Eng - 26, 200*.

Barratt, Samuel, 1843-91, gas engineer.
    PIME - 1891, 606.

Barrett, Alfred,   -1872, ironworks engineer.
    Eng - 35, 6*.

Barrett, James, 1808-59, heating engineer.
    PICE - 19, 185-186.

14

Barrett, John James, 1846-97, mechanical engineer.
    PIME - 1897, 513-514.

Barrie, William, 1849-98, engineer overseas.
    PIME - 1898, 701.

Barrington, William, 1825-95, railway engineer.
    PICE - 120, 341-342.

Barrow, Maj. Knapp, 18(35)-88, telegraph engineer.
    Elec Rev - 23, 443-444.

Barry, Frederick, 1821-85, contracting engineer.
    PICE - 83, 430-432.

Barry, John David, 1832-81, railway engineer in Europe.
    PICE - 65, 379-380.

Barry, Philip, 1859-95, civil engineer.
    PICE - 125, 397.

Barry, William Henry, 1824-90, amateur engineer.
    PICE - 100, 413-414.

Bartholomew, Hugh,   -1885, gas engineer.
    Engng - 39, 189*.

Bartlett, Thomas, 1818-64, railway contractor(B).
    PICE - 24, 526-527.

Barton, Edwin, 1831-77, railway engineer.
    PICE - 49, 265-266.

Bartram, George, 18(36)-91, shipbuilder.
    Iron - 38, 164.

Barwell, William Harrison, 1822-64, foundry owner.
    PIME - 1865, 13.

Bassett, Alexander, 1824-87, civil & mining engineer.
    PICE - 92, 377-379.

Bateman, Frederic Foster la Trobe, 1853-89, consulting engr.
    PICE - 96, 316-317.

Bateman, John Frederic la Trobe, F.R.S., 1810-89, water eng-
    ineer(D).
    Elec - 23, 137.
    Eng - 67, 509.
    Engng - 47, 672-673.
    Iron - 33, 515.
    PICE - 97, 392-398.

Bateman-Champain, Col. Sir John Underwood, 1835-87, Royal
    Engineer(D) .
    Elec - 18, 283,332.
    Elec Rev - 20, 114.
    JIEE - 16, 79-91

Bates, Edward,   -1887, gas engineer.
    PICE - 90, 439.

Batho, William Fothergill, 1828-86, consulting civil engineer.
    Engng - 41, 556.
    JISI - 29, 804-805.
    PICE - 86, 353.
    PIME - 1886, 261-262.

Batten, James Brend, 1830-97, railway agent(B).
    PICE - 130, 320.

Batterbee, Ranson Colecome, 1838-88, railway engr in Brazil.
    PICE - 95, 383-384.

Battersby, James, 1843-77, engineer in India.
    PICE - 50, 185-186.

Battle, Arthur Edwin, 1857-94, commercial traveller.
    PIME - 1894, 160.

Bawden, William, 1824-88, mechanical engineer.
    PIME - 1889, 193.

Baxendale, Joseph, 1785-1872, railway chairman(B).
    PICE - 36, 284-285.

Bayley, Henry, 1826-87, shipping company director.
    PICE - 90, 444-446.

Baylis, Henry, 1824-74, municipal engineer.
    PICE - 40, 245.

Bayliss, John, 1826-1900, railway contractor.
    PICE - 143, 338-339.

Bayliss, Moses, 1816-94, ironmaster.
    JISI - 47, 260.

Bayliss, Samuel, 1821-98, civil engineer overseas.
    PICE - 136, 362-363.

Bayly, George Henry, 1834-89, railway engineer overseas.
    PICE - 93, 387-388.

Bayly, George Henry, 1845-87, Public Works engr in India.
    PICE - 93, 478-479.

Bazalgette, Sir Joseph William, 1819-91, municipal engineer
(B,D).
Eng - 71, 231, 354.
Engng - 51, 342-343.
Iron - 37, 253.
PICE - 105, 302-308.
PIMCE - 17, 289*.

Beamish, Richard, F.R.S., 1798-1873, railway engineer(B).
PICE - 40, 246-251.

Beardmore, Nathaniel, 1816-72, hydraulic engineer(B,D).
PICE - 36, 256-264.

Beardmore, Nathaniel St.Bernard, 1848-85, engr in India.
PICE - 82, 378-380.

Beardmore, William, 1824-77, inventor & patentee(B).
JISI - 12, 292-293.
PICE - 51, 268-270.

Beattie, Francis Henry, 1843-87, civil engineer.
PIME - 1888, 153-154.

Beattie, Joseph Hamilton, 1808-71, locomotive engineer(B).
Engng - 12, 250*.
PICE - 33, 204-206.
PIME - 1872, 15-16.

Beattie, William, 1836-82, locomotive engineer.
PICE - 72, 309-310.

Beatty, James, 1820-56, railway engineer(B).
PICE - 16, 154, 158.

Beauchamp, Lt.-Col. Clayton Scudamore, 1842-89, Royal Engr.
PICE - 100, 414-415.

Beaufort, Rear-Adm. Sir Francis, F.R.S., 1774-1857, hydrogr-
apher(D).
PICE - 18, 186-188.

Becher, Harry Macdonald, 1855-93, mining engineer overseas.
PICE - 115, 397-398.

Becher, Septimus J   , 18(64)-98, surveyor.
Eng - 87, 174.

Beckett, George, 18(44)-93, mechanical   engineer.
TMAE - 3, 286*.

Beckwith, John Henry, 1839-98, engineering manufacturer.
Elec Rev - 41, 891.
Eng - 85, 577.

Engng - 65, 761.
JISI - 54, 325.
PIME - 1898, 701-702.

Beddoe, E   ,   -1896, colliery owner.
Eng - 82, 352*.

Bedford, John, 18(15)-98, ironmaster.
Eng - 86, 431.
Engng - 66, 554.

Bedlington, Peter Roe, 1856-93, mining engineer overseas.
PICE - 113, 351-352.

Bedson, George, 1820-84, ironworks manager(B).
JISI - 27, 539-540.

Beit, H   H   , 18(73)-99, electrical engineer.
Elec Rev - 44, 23*.
Eng - 87, 22*.

Beith, William,   -1898, colliery owner.
Eng - 86, 46*.

Belk, William, 1849-98, dock engineer.
Engng - 66, 118.
PICE - 135, 327.

Bell, Alexander Morton, 1828-85, railway engineer.
PICE - 82, 381.

Bell, Andrew,   -1896, telegraph engineer.
Elec Rev - 38, 249.

Bell, Edward, 1812-75, civil engineer.
PICE - 42, 255-258.

Bell, George, 1808-70, manufacturing engineer.
PIME - 1871, 15.

Bell, Henry Stowe,   -1894, metallurgist.
JISI - 45, 389*.

Bell, James, 1829-83, municipal surveyor.
PICE - 79, 361-362.

Bell, James, 1852-98, mining engineer.
PICE - 135, 363-364.

Bell, John, 18(20)-88, iron smelter.
Engng - 45, 97.
JISI - 32, 207-208.

Bell, John, 18(30)-91, iron founder.
  Engng - 51, 464.

Bell, John Thomas, 1809-86, iron manufacturer.
  JISI - 28, 801-802.

Bell, Philip Henry, 18(65)-95, electrical engineer in Cape
  Town.
  Elec - 36, 245.
  Elec Rev - 37, 765.

Bell, Robert, 1817-86, railway engineer.
  Engng - 42, 120*.

Bell, Dr Robert,   -1894, oil engineer.
  JISI - 45, 389*.

Bell, Robert Bruce, 18(23)-83, civil engng contractor.
  Eng - 56, 168.
  Engng - 36, 152, 191.
  PICE - 75, 293-296.
  TIESS - 27, 213.

Bell, Thomas,   -1875, railway engineer.
  PICE - 42, 261.

Bell, Thomas, 1806-94, ironmaster.
  JISI - 46, 264.

Bell, Thomas, 1810-74, waterworks engineer.
  Engng - 17, 389*.
  PICE - 41, 223.

Bell, Thomas, 1846-84, engineering manufacturer.
  JISI - 25, 553-554.

Bell, William, 1818-92, civil engineer.
  PICE - 109, 390-392.

Bell, Wilson, 1839-88, railway engineer in India.
  PICE - 101, 285-287.

Bellhouse, Edward Taylor, 1816-81, engineering manufacturer(B).
  PIME - 1882, 1-2.

Bembridge, J   K   ,   -1892, manufacturer.
  Iron -40, 451.

Benest, James Smyth, 1826-96, municipal engineer overseas.
  PICE - 125, 416.

Bennett, Clifton George,   -1899, municipal surveyor.
  PIMCE - 26, 253*.

Bennett, John, 1823-96, Public Works engineer overseas.
PICE - 128, 341-342.

Bennett, Peter Duckworth, 1825-85, engineering manufacturer.
JISI - 27, 540-541.
PICE - 83, 432.
PIME - 1885, 525.

Bennett, Samuel, -1859, mechanical engineer overseas.
PICE - 19, 186*.

Bennett, William Christopher, 1824-89, civil engr overseas.
PICE - 99, 346-350.

Bennett, William Henry, 1828-92, gas engineer.
TSE - 1892, 244.

Benson, Sir John, 1812-74, civil engineer & surveyor(B,D).
Eng - 38, 313*.
PICE - 40, 251-253.

Benson, William Robert, -1897, colliery owner.
Eng - 83, 203*.

Bentall, Edward Hammond, 1814-98, agricultural implement man-
ufacturer(B).
Eng - 86, 157*.

Berbeck, G , -18(84), consulting engineer.
Eng - 38, 7*.

Berkely, Sir George, 1821-93, railway consulting engineer(B).
Elec - 32, 211*.
Eng - 76, 617.
Engng - 56, 796.
JISI - 45, 389-391.
PICE - 115, 382.

Berkely, James John, 1819-62, railway engineer in India(B).
Eng - 14, 156.
PICE - 22, 618-624.

Bernays, Edwin Arthur, 1822-87, dock engineer.
PICE - 91, 408-411.

Berrey, Thomas Hope, 18(16)-94, waterworks engineer.
PICE - 116, 386-387.

Berryman, Robert, 18(38)-77, inventor.
Engng - 24, 208*.

Berthon, Revd Edward Lyon, 1813-99, inventor(B,D).
Eng - 88, 453+port.
Engng - 68, 563.

Bertram, Alexander, 1853-99, mechanical engineer.
    PIME - 1899, 127.

Bessemer, Sir Henry, F.R.S., 1813-98, steel manufacturer(B,D).
    Elec - 40, 693-694.
    Elec Rev - 41, 372.
    Eng - 85, 256, 283+port.
    Engng - 65, 341-343+port.
    JISI - 53, 298-311.
    PICE - 134, 394-399.
    PIME - 1898, 133-136.
    Prac Mag - 6, 97-103+port.
    TIESS - 41, 377-378.
    TSE - 1898, 234-236.

Best,Stephen Charles, 1831-91, railway engineer.
    PICE - 109, 392-393.

Bestic, William Brereton, 1859-92, engineer in India.
    PICE - 114, 387-388.

Bethell, John, 1804-67, patentee(B).
    Mech Mag - 86, 142-143.
    PICE - 27, 597-599.

Betts, Edward Ladd, 1815-72, railway contracting engineer(B).
    Engng - 13, 157.
    PICE - 36, 285-288.

Betts, James Alexander, 1845-92, telegraph engineer(B).
    Elec Rev - 31, 234.
    PICE - 111, 398-399.

Bevan, Isaiah,1851-92, steel manufacturer.
    Iron - 42, 97*.
    JISI - 41, 301-302.

Bevan, Morgan Mark , 1849-99, telegraph engineer overseas.
    JIEE - 29, 948.

Bevan, William,   -1897, steel manufacturer.
    Eng - 84, 144*.
    JISI - 53, 312*.

Bewick, Thomas John, 1821-97, mining engineer.
    PICE - 131, 361-363.
    TFIME - 15, 172-175+port. frontis. By T.Burrell Bewick.

Bewley, Thomas Arthur, 1844-89, shipbuilder.
    PIME - 1889, 746-747.

Beyer, Charles Frederick, 1813-76, railway loco. manufacturer(B).
    Eng - 41, 439.

Engng - 21, 492*, 505.
PICE - 47, 290-297.
PIME - 1877, 16-17.

Bickle, Thomas Edwin, 1857-98, mining engineer.
PIME - 1898, 308.

Bicknell, Edward, 1849-98, mechanical engineer overseas.
PIME - 1898, 308-309.

Bidder, George Parker, 1806-78, civil engineer(B,D).
Eng - 46, 228-229.
Engng - 26, 260.
Iron - 12, 396.
PICE - 57, 294-309.

Bidder, George Parker, Q.C., 1836-96, railway lawyer(B).
Engng - 61, 202.
PICE - 125, 422-428.

Bidder, Samuel Parker, 1843-78, mining engineer(B).
PICE - 53, 284-285.

Biddle, J   P   , 18(41)-92, ironmaster.
Iron - 40, 384*.

Bigg, Henry Heather, 1826-81, instrument manufacturer(B).
PICE - 68, 317-320.

Bigg-Wither, Thomas Plantagenet, 1845-90, rly engr overseas.
PICE - 103, 381-383.

Bigsby, Lt. Gordon, 18(38)-73, Royal Engineer.
PICE - 39, 285-286.

Bingham, Charles Henry, 1848-1900, manufacturer.
Eng - 90, 351.
PIME - 1900, 622.

Binns, Charles, 1813-87, railway engineer.
Eng - 63, 67.
TCMCIE - 16, 9-11.

Binns, William, 1815-95, consulting mechanical engineer.
PICE - 123, 441.

Birch, Eugenius, 1818-84, contractor.
PICE - 78, 414-416.

Birch, John Brannis, 1813-62, civil engineer.
PICE - 22, 640-641.

Birch, John Grant, 1847-1900, contractor.
Eng - 90, 115*, 116+port.

Engng - 70, 153.
PIME - 1900, 623.

Birch, Robert William Peregrine, 1845-96, consulting sanitary engr.
Elec Rev - 39, 341.
Engng - 62, 339.
PICE - 127, 362-364.
PIME - 1896, 596-597.
TSE - 1896, 241-242.

Bird, Thomas,   -1895, works manager.
Elec - 34, 388*.
Elec Rev - 36, 126*.
Eng - 79, 113.

Bird, William, 1805-84, ironmaster.
Eng - 37, 139*.
Iron - 23, 98.

Birkinshaw, John Cass, 1811-67, railway engineer(B).
PICE - 31, 202-207.

Birtwistle, Richard, 1834-92, engineering manufacturer.
PIME - 1892, 223.
TMAE - 2, 277.

Bishop, Arthur William, 1868-98, civil engineer.
PICE - 133, 413-414.

Bittleston, W   H   ,   -1891, naval architect.
NECIES - 7, xlix.

Black, Frank Charles, 1839-89, Public Works engr in India.
PICE - 98, 402-403.

Black, George, 1823-73, civil engineer.
PICE - 40, 253-255.

Black, James Mark, 1852-98, mechanical engineer.
PIME - 1898, 528.

Blackadder, William, 1789-1860, agricultural engineer.
PICE - 20, 136-137.

Blackburn, James, 1835-82, gas engineer.
PICE - 70, 413-415.

Blackburne, John George, 1815-71, surveyor(B).
PICE - 33, 206-209.

Blackburne, John William, 1839-74, contracting engineer.
PICE - 41, 210-211.

Blackett, John, 1819-93, consulting engr in New Zealand.
PICE - 113, 331-332.

Blackett, John George, 1852-85, civil engineer.
PICE - 83, 438-439.

Blackwell, Samuel Holden, 1816-68, coal & ironmaster.
PIME - 1869, 15-16.

Blackwell, Thomas Evans, 1819-63, railway engineer(B).
PICE - 23, 481-485.

Blackwood, Capt. Alexander,   -18(83), shipbuilder.
TIESS - 27, 214-215.

Blackwood, Thomas, 1819-91, shipbuilder.
Eng - 72, 419.
Engng - 52, 596-597.
Iron - 38, 454.
TIESS - 35, 301.

Bladen, Charles, 1829-72, steel works manager.
PIME - 1873, 17.

Blair, James Fairlie, 1831-76, railway engineer.
Engng - 22, 127-128.
PICE - 49, 258-259.

Blair, Peter Robert, 1866-91, electric light contractor.
PIME - 1891, 189.

Blair, Thomas, 1854-82, gas engineer.
PICE - 69, 416.

Blair, William Newsham, 1841-91, civil engr & surveyor in
New Zealand.
Engng - 52, 104*.
PICE - 107, 396-398.

Blake, Henry Wollaston, F.R.S., 1815-99, engineering manufr(B).
Eng - 88, 5.
Engng - 68, 30.
PICE - 138, 488-489.
PIME - 1899, 467-470.

Blake, Thomas,   -1892, iron merchant.
Iron - 39, 449.

Blakelock, Percy,   -1898, waterworks engineer.
Eng - 85, 399.

Bland, Thomas, 18(19)-87, gun manufacturer.
Iron - 29, 343.

Blatchley, Charles Graham, 1828-87, railway engineer.
PICE - 91, 411-412.

Blechynden, Alfred, 1849-97, marine engineer.
Eng - 83, 213.
Engng - 63, 287.
PIME - 1897, 131-132.
TINA - 38, 314.

Bleckly, Henry, 1812-90, coal and ironmaster(B).
JISI - 36, 175-176.

Bloor, James,   -1896, colliery engineer.
TIME - 10, 163 -164.

Blyth, Benjamin Hall, 1819-66, railway engineer(B).
PICE - 26, 556-560.

Bocquet, William Sutton, 1848-89, railway engineer in India.
PIME - 1889, 747.

Bodden, George, 1850-86, textile machinery manufacturer.
PIME - 1886, 461-462.

Bodmer, John George, 1786-1864, mechanical engr, inventor.
PICE - 28, 573-608.

Bölckow, Henry William Ferdinand, M.P., 1806-78, ironmaster
(B,D).
Eng - 45, 444.
Engng - 25, 494.
Iron - 11, 778-779.
JISI - 12, 228-230.
PICE - 55, 325-326.

Bold, Edward, 1841-1900, telegraph engineer in New Zealand.
JIEE - 30, 1244.
PICE - 142, 385-386.

Bolden, Henry, 1837-91, railway engineer.
PICE - 109, 393-394.

Bolland, James, 1835-85, civil engineer in Brazil.
PICE - 81, 323-324.

Bolton, Col. Sir Francis John, 1830-87, Royal Engineer, elec-
trician(B,D).
Elec - 18, 190.
Elec Rev - 20, 37.
Eng - 63, 36.
Iron - 29, 31.
JIEE - 16, 2-4.
PICE - 93, 497-501.

Bond, George, 1840-96, coal & ironworks manager.
   JISI -49 , 283-284*.
   PICE - 125, 397-398.

Bone, Charles Christianson, 1844-92, water engineer.
   PICE - 111, 364-366.
   TIESS - 35, 301-302.

Bonnell, John,   -1882, agricultural machine designer.
   Eng - 54, 25*.

Booker-Blakemore, Thomas William, M.P., 1801-58, tinplate
   manufacturer(B).
   PICE - 18, 202-203.

Booth, William Stanway, 1853-94, mechanical engineer.
   PIME - 1895, 140.

Boothby, Alexander Cunningham, 1857-88, consulting engineer.
   PICE - 97, 420-421.

Borrie, John, 1837-84, consulting engineer(B).
   PIME   - 1884, 398.

Borrows, William, 1850-1900, manufacturer.
   PIME - 1900, 623-624.

Borthwick, Michael Andrews, 1810-56, railway engineer(B).
   PICE - 16, 108-113.

Bottomley, William Fereday, 18(40)-91, telegraph engineer.
   Elec - 28, 4.
   Elec Rev - 29, 531-532.

Bouch, Sir Thomas, 1822-80, consulting railway engineer(B,D).
   Eng - 50, 347-348.
   Engng - 30, 409.
   PICE - 63, 301-308.
   PIME - 1881, 1.

Bouch, William, 1813-76, locomotive manufacturer.
   Eng - 41, 63.
   PIME - 1877, 17.

Boulton, Isaac Watt, 18(24)-99, locomotive manufacturer.
   Eng - 87, 620.

Boulton, Thomas Fletcher,   -1896, telegraph engineer.
   Elec Rev - 38, 85.

Bourdeaux, John, 1834-90, electrical engineer.
   Elec - 25, 229.
   PICE - 104, 308-309.

Bourke, H   G   ,   -1898, admiralty engineer.
     Eng - 85, 233.

Bourne, John Frederick, 1816-79, civil engineer overseas.
     PICE - 59, 289-291.

Bourne, Joseph, 1836-93, locomotive superintendent.
     PICE - 114, 367.

Bovill, George Hinton, 1821-68, engineering manufacturer(B).
     PIME - 1869, 16.

Bower, Anthony, 1820-91, foundry owner.
     PIME - 1891, 189-190.

Bown, William, 1834-1900, manufacturer.
     Eng - 90, 114.

Bowser, Howard, 1824-99, pipe manufacturer.
     Eng - 88, 275.
     Engng - 68, 337.
     JISI - 56, 292.
     TIESS - 43, 354.

Box, Thomas, 18(21)-85, engineering writer.
     Eng - 60, 124.

Boxer, Capt.Edward Mourrier, F.R.S., 1822-98, Royal Artillery
     (B).
     Eng - 85, 8-9.

Boyd, Edward Fenwick, 18(09)-89, coal owner.
     Eng - 68, 245.
     Iron - 34, 228.
     TIME - 1, 76, 221; 2, 204.
     PICE - 89, 455.

Boyd, John, 1810-78, engineering manufacturer.
     PICE - 55, 326-327.

Boydell, James,   -1859, mechanical engineer.
     Eng - 8, 465*.

Boys, William John, 1840-82, municipal engineer.
     PICE - 72, 310.
     PIMCE - 9, 205-206.

Brace, William Henry, 1857-92, Public Works engr in India.
     PICE - 119, 401.

Bradbury, Augustus Beattie, 1841-75, engineer in India.
     PICE - 44, 224-225.

Bradford, Hugh Meller, 1837-86, railway engr & surveyor.
PICE - 89, 456.

Bradley, Thomas, 18(23)-97, ironmaster.
Eng - 84, 610.

Brady, Joseph Henry, 1851-94, railway engineer overseas.
PICE - 119, 389-390.

Brady, Patrick,   -1872, railway engineer in Victoria.
Engng - 14, 356.

Bragge, William, 1823-84, steel manufacturer(B,D).
Eng - 57, 441.
Engng - 37, 538*.
Iron - 23, 523.
PIME - 1884, 398-399.

Braidwood, James, 1800-61, Superintendent, London Fire
Brigade(B,D).
PICE - 21, 571-578.

Braidwood, John, 18(08)-78, consulting mechanical engineer.
Engng - 25, 227.

Brain, Carl Thomas Blanch, 18(66)-95, electrical engineer.
Elec - 34, 388*.
Elec Rev - 36, 126.

Braithwaite, Frederick, 1798-1865, manufacturing engineer.
PICE - 26, 560-561.

Braithwaite, John, 1797-1870, railway engineer(B).
Eng - 30, 237*.
Engng - 10, 248*.
Mech Mag -13, 235-237, 377-388, 417-419.
PICE - 31, 207-211.

Bramah, Francis,   -1840, ---
CEAJ - 4, 127.
PICE - 1, 14.

Bramley-Moore, John, 18(00)-86, Chairman,Liverpool Docks.
Eng - 62, 431.

Brand, Charles, 1804-85, contractor.
Iron - 25, 78.

Brand, David Jollie, 1852-99, marine engineer.
PIME - 1899, 470.

Brandreth, Lt.-Col.Henry Rowland, F.R.S., 1794-1848, Royal
Engineer.

CEAJ - 11, 96*.
PICE - 8, 12-15.

Brandt, John, 1838-86, railway engineer.
PICE - 88, 432-433.

Brassey, Thomas, 1805-70, railway contractor(B,D).
Eng - 30, 406.
Engng - 10, 443.
PICE - 33, 246-251.

Bratt, Augustus Hicks Henery, 1863-96, consulting engineer in
China.
JISI - 53, 313.
PIME - 1897, 132.

Bray, William Bayley, 1811-85, engineer in New Zealand.
PICE - 84, 439-441.

Brayshaw, Thomas H   , 1850-93, telegraph engineer.
Elec - 30, 673.

Brebner, Alan, 1826-90, lighthouse contractor.
Engng - 49, 335.
PICE - 101, 287-289.

Brebner, Samuel Gordon, 1848-95, armaments engineer.
PIME - 1895, 308-309.

Breeden, Joseph, 1821-93, inventor, mechanical engineer.
PIME - 1893, 490.

Bremner, Alexander,   -1862, civil engineer.
PICE - 23, 505-506.

Bremner, David, 1818-52, Glasgow Harbour Engineer.
PICE - 12, 148-149.

Bremner, James, 1784-1856, shipbuilder(B,D).
CEAJ - 19, 322*.
Eng - 2, 466.
PICE - 16, 113-120.

Brentnall, William, 1829-94, waterworks engineer.
PICE - 118, 442-443.
PIMCE - 20, 333; 21, 323-324.

Brereton, William,   -1899, Secretary, Amalgamated Society
of Engineers.
Eng - 88, 635.

Brett, Jacob, 1808-97, telegraph engineer.
Elec - 38, 368+port.
Elec Rev - 40, 75.

Brewer, John Williams, 1841-94, railway engineer.
    Engng - 58, 297.
    PICE - 119, 402-403.

Brewster, Edward, 1860-97, civil engineer(B).
    PICE - 132, 394*.

Brickenden, James Gordon, 1848-87, railway engineer.
    PICE - 92, 382-383.

Bridgeman, Henry Orlando, 1825-79, railway engineer.
    PICE - 58, 339-340.

Bridges, John Douglas Ormond, 1866-95, railway engineer
    overseas.
    PICE - 122, 394.

Bridgford, Maj. Sidney Thomas, 1836-97, Royal Artillery.
    PICE - 133, 414-415.

Brierley, Richard, 18(38)-99, municipal engineer.
    PIMCE - 26, 253.

Briggs, Henry Currer, 1829-81, ironmaster & chemist.
    JISI - 21, 659-661.

Bright, Sir Charles Tilston, 1832-88, telegraph engineer(B,D).
    Elec - 21, 13.
    Elec Rev - 23, 508-512.
    Eng - 65, 387.
    Iron - 31, 421.
    JIEE - 17, 477-478.
    PICE - 93, 479-487.

Bright, Hugh Meyler, 1840-66, ---
    Eng - 21, 52*.

Bright, James,   -1893, iron works manager.
    Iron - 41, 143*.

Briscoe, Capt. Alfred P   ,   -1890,cable ship captain.
    Elec - 24, 488*.

Bristow, Henry William, F.R.S., 1817-89,   geologist.
    Iron - 33, 539.

Broad, Robert, 1821-74, ironmaster.
    PIME - 1875, 19*.

Broadbent, Thomas, 1833-80, mechanical engineer.
    PIME - 1881, 2.

Broadberry, William Henry Hague, 1831-99, gas works manager.
    TSE - 1899, 260-261.

Brock, Aubrey, 1842-95, brick manufacturer.
    JISI - 49, 284*.

Brockat, John, 1835-87, marine surveyor.
    NECIES - 3, xxxvi.

Brockbank, William, 1829-96, railway surveyor.
    JISI - 50, 257*.

Brockedon, Philip North, 1822-49, mechanical engineer.
    PICE - 10, 95-97.

Brogden, Alexander, 18(16)-92, ironmaster.
    Iron - 40, 495.

Brogden, George W   H   ,   -1892, colliery owner.
    Iron - 40, 451.

Brogden, John, 1823-55, contractor.
    PICE - 15, 94-96.

Bromley, Edmund, 1808-92, colliery engineer.
    TIME - 5, 480.

Bromley, Massey, 1846-84, consulting engineer.
    Eng - 58, 130.
    Engng - 38, 94.
    Iron - 24, 119.
    JISI - 25, 561-562.
    PICE - 82, 382-383.
    PIME - 1884, 400.

Brook, Thomas, 1818-86, surveyor.
    PICE - 89, 490-491.

Brooke, Sir William O'Shaughnessy, 1809-89, telegraph engr.
    Elec - 22, 297.
    Elec Rev - 24, 68.

Brookes, William, 1817-84, patent agent.
    Eng - 57, 155.
    PICE - 76, 372.

Brooks, William Alexander, 1802-77, civil engineer(B).
    PICE - 50, 172-175.

Brotherhood, Arthur Maudslay, 1867-93, railway engineer.
    PIME - 1893, 491.

Brotherhood, Rowland, 18(13)-83, railway contractor.
    Engng - 35, 219.
    Iron - 21, 230.

Brothers, Horatio, 1822-99, consulting gas engineer.
    PICE - 141, 336.

Brough, Lionel,   -1876, Inspector of Mines.
   Engng - 22, 232 .
   JISI - 9, 476*.

Brough, Richard Secker, 1846-79, telegraph engineer in
   India(B).
   JSTE - 8, 281*, 397-399. By Louis Schwendlar.
   PICE - 59, 315-317.

Brown, Charles Edwin, 1856-1900, surveyor.
   PICE - 144, 312-314.

Brown, Col. Francis David Millett, V.C., 1837-95, Public
   Works engineer in India.
   PICE - 124, 437-438.

Brown, George, 1832-74, engineering manufacturer.
   Eng - 38, 250*.
   Engng - 18, 273*.

Brown, Henry, 1832-91, railway stores contractor.
   PIME - 1892, 98.

Brown, James, 1790-1872, mechanical engineer.
   Eng - 33, 207.

Brown, James,   -1898, civil engineer.
   Engng - 66, 326.

Brown, John, 1815-82, estate agent.
   TCMCIE - 11, 14-15.

Brown, Sir John, 1816-96, steel manufacturer(B,D).
   Eng - 83, 5+port.
   Engng - 63, 23.
   Prac Mag - 7, 33-36.

Brown, John, 1823-88, mining engineer.
   JISI - 33, 167-169.
   PICE - 95, 361-363.
   TCMCIE - 17, 263-266.

Brown, Samuel, 1836-91, civil engineer overseas.
   Iron - 38, 493.
   PICE - 109, 395-398.

Brown, Thomas, 1812-87, naval Inspector of Machinery.
   PICE - 92, 393-394.

Brown, Thomas John, 18(45)-91, telegraph engineer.
   Elec - 27, 599.

Brown, Capt. William,   -1884, ---
   TIESS - 28, 290*.

Brown, William Armitage, 1836-88, railway engineer.
PICE - 95, 363-364.

Brown, William Joseph, 1835-94, engineering works manager.
PICE - 119, 403-404.

Brown, William Steel, 1835-64, locomotive superintendent.
PIME - 1865, 13.

Browne, Andrew, 1857-1900, marine superintendent.
Eng - 89, 356.
Engng - 69, 445.

Browne, Edward Fiske,1820-50, railway surveyor.
PICE - 10, 97.

Browne, Maj.-Gen. Sir James, 1839-96, Royal Engineer(B).
PICE - 125, 428-430.

Browne, Valentine, 1824-83, railway engineer.
PICE - 74, 283.

Browne, Walter Raleigh, 1842-84, engineering manufacturer(B).
Eng - 58, 205.
Iron - 24, 273.
PICE - 79, 362-366.
PIME - 1884, 472.

Browning, Arrott, 1838-77, railway engineer.
PICE - 51, 265-266.

Browning, Thomas Gaul, 1831-73, municipal surveyor.
PICE - 39, 286.

Brownlee, James, 1813-90, manufacturer.
Engng - 50, 496.
TIESS - 34, 320-321.

Brownlee, Walter Hughes, 18(68)-97, electrical engineer.
Elec - 40, 313*.
Elec Rev - 41, 905*.

Brown-Westhead, T   C   , 1837-82, manufacturer.
JISI - 23, 666.

Bruce, William Duff, 1839-1900, engineer in India.
Engng - 69, 591.
JISI - 57, 251.
PICE - 141, 336-339.
PIME - 1900, 323-325.

Bruff, Peter Schuyler, 1812-1900, civil engineer(B).

Eng - 89, 224.
PICE - 141, 339-341.

Brundell, Benjamin Shaw, 1825-97, civil engineer.
PICE - 128, 343-344.

Brunel, Isambard Kingdom, F.R.S., 1806-59, civil engr(B,D).
Eng - 8, 219, 326, 380, 385, 396.
PICE - 19, 169-173.

Brunel, Sir Mark Isambard, F.R.S., 1769-1849, civil engineer(B,D).
CEAJ - 13, 18-19.
PICE - 10, 78-81.

Brunlees, Sir James, 1816-92, contractor, rly & dock engr(B,D).
Eng - 73, 503.
Engng - 53, 729-730.
Iron - 39, 516.
JISI - 41, 298-299.
PICE - 111, 367-371.
PIME - 1892, 223-224.

Brunton, George, 1823-1900, ironworks manager in India.
Eng - 89, 357*.
PICE - 141,341*.

Brunton, John, 1812-99, railway contracting engineer.
PICE - 136, 345.

Brunton, Robert, 1796-1852, ironworks engineer(B).
PICE - 12, 149-51。

Brunton, Thomas,   -1894, railway manager.
Eng - 77, 245.

Brunton, William, 1777-1851, civil & mechanical engineer(B).
PICE - 11, 95-99.

Brunton, William, Jun., 1817-81, railway & mining engineer in India(B)
PICE - 67, 395-396.

Brunton, William Alexander, 1840-81, rly engineer in India.
PICE - 64, 338-339.

Bryce-Douglas, Archibald Douglas, 1839-91, shipbuilder & marine
engineer.
Eng - 71, 287.
Engng - 51, 429.
Iron - 37, 315.
PIME - 1891, 190-191.
TINA - 132, 352-353.

Bryham, William, 18(17)-93, mining engineer.
Iron - 41, 316.

Buchan, John,   -1895, municipal engineer.
    Eng - 79, 322.

Buchanan, George, 1827-97, railway engineer(B).
    Eng - 83, 602.
    PICE - 129, 363.

Buchanan, William M   ,   -1856, shipbuilder.
    Eng - 2, 384*.

Bucholz, John Augustus Arnold, 1846-89, flour manufacturer.
    PICE - 101, 289-292.

Buck, George Watson, 1789-1854, railway engineer.
    PICE - 14, 128-130.

Buck, Joseph Hayward Watson, 1839-98, railway engineer.
    PICE - 134, 402.

Buckham, Thomas, 1836-77, sanitary engr & surveyor.
    PICE - 53, 285-286.

Buckley, James, 1838-95, iron & tinplate manufacturer.
    JISI - 48, 350*.

Buckney, Thomas, 1838-1900, chronometer manufacturer(B).
    JIEE - 29, 948.
    PIME - 1900, 326.

Budd, James Palmer, 1803-83, ironmaster.
    Eng - 56, 471.

Buddicom, William Barber, 1816-87, railway contractor(B).
    Eng - 64, 180.
    Engng - 44, 170.
    PICE - 91, 412-421.
    PIME - 1887, 466-467; 1888, 154.

Buddle, John, 1773-1843, colliery viewer(D).
    PICE - 3, 12-13.

Bulkeley, Capt. Thomas, 1807-82, railway director.
    Engng - 33, 504*.

Bullmore, Frederick Charles, 1842-87, railway engr in India.
    PICE - 92, 383-384.

Bunning, Theophilus Wood, 1822-88, Secy, Northumberland & Durham
    Coal Trades Associations.
    Engng - 46, 61*.
    TIME - 3, 85-87. By C.Z. Bunning.

Burgess, John, 1825-94, gas engineer.
    PICE - 119, 404-405.

Burgess, William Edward, 1868-96, mechanical engineer.
PICE - 129, 392.

Burgon, Frederick, 18(61)-1900, manufacturer.
Eng - 89, 265.

Burgoyne, Gen. Sir John Fox, Bart., 1782-1871, Royal Engr(B,D).
Eng - 32, 248-249.
PICE - 33, 192-203.

Burleigh, Benjamin, 1820-76, railway engineer.
PICE - 47, 301-302.

Burlinson, William Davie, 1802-61, manufacturer.
PIME - 1862, 14.

Burn, James, 18(32)-97, mining engr & iron merchant.
Eng - 83, 401.

Burnell, George Rowdon, 1814-68, hydraulic engineer(B).
PICE - 31, 211-212.

Burnet, Lindsay, 1855-95, boiler manufacturer.
Eng - 79, 239.
Engng - 59, 374.
PICE - 122, 395-396.
PIME - 1895, 140-141.
TIESS - 38, 331-332.
TMAE - 5, 264-265.

Burnett, Arthur Wildman, 1844-90, hydraulic engineer in Ceylon.
PICE - 103, 360-362.

Burnett, Robert Reginald, 1841-83, railway & mining engineer.
PICE - 75, 297-298.

Burns, Andrew,   -1885, shipyard manager.
TIESS - 29, 219.

Burns, Sir George, Bart., 1795-1890, ship owner(B,D).
Eng - 69, 458.
Engng - 49, 682.

Burns, Jerome, 1827-94, railway engineer in India.
PICE - 117, 363-364.

Burns-Lindow, Jonas Lindow,   -1893, ironworks manager.
JISI - 43, 290.

Burrell, Percy, 1833-90, railway engineer in South America.
PICE - 103, 362-363.

Burridge, Stephen, 1823-88, steelworks manager.
JISI - 33, 171.

36

Burrows, John Henry, 18(45)-99, ironmaster.
    Eng – 84, 423.
    Engng - 67, 548*.

Burrows, S   H   , 18(17)-92, ironmaster.
    Iron - 40, 540.

Burstal, Capt. Edward, R.N., 1818-86, marine & telegraph engr(B).
    Eng - 62, 76.
    Engng - 42, 81-82.
    Iron - 28, 88.

Burstall,Timothy, 1776-1860, inventor.
    Eng - 10, 391*.

Burt, J   E   , 18(52)-93, telegraph engineer overseas.
    Elec - 30, 58.

Burtchaell, Peter, 1820-94, municipal engineer.
    PIMCE - 21, 325.

Burton, H   M   , 18(11)-66, ---
    Eng - 21, 203*.

Burton, Prof.William Kinninmond, 1856-99, sanitary engr in Japan.
    Eng - 88, 634. Port. in v.84, 567.
    PICE – 139, 373-374.

Bury, William Tarleton, 1835-76, steel manufacturer.
    PIME - 1877, 17-18.

Bushell, Henry,   -1897, agricultural machinery manufacturer.
    Eng - 84, 302.

Butler, Ambrose Edmund, 1816-83, forge master.
    PIME - 1884, 61-62.

Butler, John, 1822-84, ironmaster.
    PIME - 1885, 71-72.

Butler, John Octavius, 1812-83, forge owner.
    Iron - 22, 379.
    PICE - 75, 298.

Butler, Richard, 1822-99, railway engineer & surveyor.
    PICE - 137, 418.

Butler, Thomas Snowden, 1846-74, engineering works manager.
    PIME - 1875, 19.

Butlin, William, 18(24)- 97, ironmaster.
    Eng - 83, 162.
    JISI - 51, 309-310.

Butter, Frederick Henry, 1847-99, mechanical engineer.
PIME - 1900, 326-327.

Butterton, William, 1854-99, civil engineer & contractor.
PICE - 140, 267.

Butterworth, George, 1839-98, mining engineer.
TFIME - 16, 127.

Byrne, Oliver, 18(10)-80, inventor.
Eng - 50, 474.

Cabry, Joseph,   -1897, railway engineer.
Eng - 84, 436*.

Cabry, Thomas, 1801-73, railway engineer.
PIME - 1874, 16-17.

Cadell, Henry, 1812-88, ironmaster.
JISI - 33, 169- 170.

Cadett, William James, 1847-76, civil engineer.
PICE - 51, 270-271.

Cail, Richard, 1812-93, contractor.
PICE - 115, 400-402.

Caine, Nathaniel,   -1877, iron merchant.
JISI - 11, 540*.

Caird, James Tennent, 1816-88, shipbuilder(B).
Eng - 65, 116.
Engng - 45, 122-123.
Iron - 31, 122.

Cairns, Andrew Duncan, 1850-96, dock & harbour engineer.
PICE - 126, 391.

Calder, Augustus, 18(53)-96, telegraph engineer.
Elec - 36, 470.
Elec Rev - 38, 177.

Caley, James Augustus, 1824-85, sanitary engineer.
PICE - 80, 330-331.

Calver, Capt. Edward Killwick, R.N., F.R.S., 1813-92, hydrographer.
PICE - 112, 373-374.

Calvert, George Alexander, -1896, marine superintendent(B).
TINA - 38, 315.

Calvert, Thomas, 18(42)-96, manufacturer.
Eng - 82, 360.
Engng - 62, 461.

Cameron, Augustus John Darling, 1841-84, tramway engineer.
PICE - 80, 331.

Cameron, John, 18(18)-80, ---
Eng - 49, 376*.

Cameron, John, 1857-91, civil engineer in India.
PICE - 109, 413-414.

Cameron, John Carter, 1840-1900, works manager.
TIESS - 44, 339.

Cammell, Charles, 1810-79, steel manufacturer(B).
JISI - 15, 615-616.
PICE - 56, 288-289.
PIME - 1880, 1-2.

Campbell, Crawford James, 1832-76, railway engineer in India.
PICE - 45, 241-242.

Campbell, David, 1813-82, consulting engineer(B).
PIME - 1883, 15.

Campbell, Dugald, 18(19)-82, sanitary engr & chemist.
Eng - 53, 365.

Campbell, George, 18(46)-89, locomotive manufacturer.
TSE - 1889, 219.

Campbell, George Stephenson, 1848-94, civil engineer.
PICE - 120, 370.

Campbell, James, 1812-84, railway & colliery engineer.
Engng - 37, 379.
Iron - 23, 399.

Campbell, Joseph C   , 18(48)-1900, engineering manufacturer.
TIESS - 43, 354-355.

Campbell, Peter Laurentz, 1809-48, railway company secretary.
PICE - 8, 14-15.

Campbell, W   R   ,   -1899, railway engineer.
Eng - 87, 33.

Cannell, Fleetwood James, 1820-71, engineering works manager.
PIME - 1872, 16.

Cannell, Thomas Stubbs, 18(41)-96, works manager.
TMAE - 6, 317.

Cantwell, Robert, 1793-1859, architect & surveyor.
PICE - 19, 186.

Capel, Herbert Churchill, 1862-99, manufacturer.
Eng - 89, 15*.
PIME - 1900, 327.

Capper, Charles, 1822-69, dock chairman(B).
PICE - 30, 465-466.

Capper, Robert, 1846-99, civil engineer.
PICE - 139, 382-383.

Cargill, John,   -1883, manufacturer.
Iron - 22, 333.

Carleton, Francis, 1800-48, shipping company director.
PICE - 8, 15-16.

Carlton, George Brody, 1854-93, sanitary engineer.
PICE - 116, 375-376.

Carlton, Thomas, 18(53)-99, Secy,Cleveland Blastfurnacemens' Assn.
Eng - 87, 43.

Carmichael, David, 1818-95, foundry owner.
Eng - 79, 398.
TIESS - 38, 332.

Carmichael, James,   -1883, forge manager.
Engng - 36, 312-313.

Carnegie, Capt.William Fullarton Lindsay, 1788-1860, Royal Artillery.
PICE - 20, 160-163.

Carpenter, William Lant, 1841-90, electrical engineer(B).
Elec - 26, 266.
Elec Rev - 28, 8.
JIEE - 20, 2.

Carpmael, Alfred, 1835-93, patent solicitor.
Elec - 30, 409*.
Eng - 75, 105.
PICE - 113, 361-362.

Carpmael, William, 1804-67, patent agent(B).
    Mech Mag - 87,43.
    PICE - 30, 430-431.
    PIME - 1868, 14-15.

Carpmael, William, 1832-99, patent agent.
    PIME - 1899, 265.

Carr, Edward, 18(36)-93, company representative.
    Iron - 41, 220.
    JISI - 43, 172*.

Carr, Henry, 1817-88, railway engineer.
    PICE - 95, 364-369.

Carr, Mark William, 1822-88, railway & mining engineer.
    JISI - 32, 209-210.
    PICE - 93, 487-488.

Carr, Robert, 1827-97, engineer, London Docks.
    Engng - 63, 519.
    PICE - 129, 364-367.
    PIME - 1897, 132-134.

Carr, Thomas, 1824-74, inventor(B).
    PIME - 1875, 19-20.

Carrack, John William, 1862-1900, works manager.
    PIME - 1900, 327-328.

Carrick, Joseph, 18(32)-87, iron merchant.
    Iron - 29, 55.

Carrick, Samuel Stewart, 1849-93, marine engineer.
    Eng - 75, 552.
    PIME - 1893, 203.

Carrington, Thomas, 1841-96, consulting mining engineer.
    PICE - 126, 392.

Carrington, W T , 18(32)-78, engineer in Singapore.
    Engng - 25, 239.

Carter, Frederick Heathcote, 1855-96, contracting engineer.
    PIME - 1896, 597.
    TMAE - 7, 282.

Carter, Richard, 1818-95, civil engineer.
    PICE - 123, 445-448.

Carter, Samuel, 1805-78, railway solicitor(B).
    JSTE - 7, 22*.

Carter, Thomas, 18(41)-94, marine engineer.
Eng - 77, 67.

Carver, Henry Clifton, 1845-93, engng manufacturer, inventor.
PIME - 1893, 491.

Case, Edward, 1842-99, consulting engineer(B).
Eng - 88, 319.
Engng - 68, 391.
PICE - 139, 374-376.

Casebourne, Charles Townshend, 1836-97, cement manufacturer.
Eng - 83, 525.
PICE - 130, 321.

Casebourne, Thomas, 1797-1864, civil engineer.
PICE - 24, 527-528.

Cassels, Robert, 18(10)-90, steel manufacturer.
Engng - 49, 178.
TIESS - 33, 209.

Castle, Henry John, 1809-91, Professor of Surveying, King's
College, London(B).
PICE - 107, 421.

Cathels, Edmund Small, 1827-83, gas engineer.
PICE - 75, 298.

Catton, John Edward, 1853-86, civil engineer in India.
PICE - 87, 446-447.

Cavendish, Lord Edward, 1838-91, industrialist(B).
JISI - 39, 231-232.
TIME - 3, 1010-1011.

Cawkwell, William, 1807-97, railway chairman(B).
PICE - 129, 398-400.

Cawley, Charles Edward, 1812-77, consulting civil engr(B).
Engng - 23, 266.
PICE - 50, 175-177.

Cayley, Sir George, Bart., 1773-1857, amateur engineer(B).
CEAJ - 21, 34.
PICE - 18, 203-204.

Cecil, Lord Sackville Arthur, 1848-98, industrialist.
Elec - 40, 487-488.
Elec Rev - 41, 152.
Eng - 85, 107.
JIEE - 28, 665-667.

Chadburn, Charles Henry,   -1890, telegraph manufacturer.
    Elec - 24, 258.

Chadwick, David, 1821-95, statistician(B).
    JISI - 47, 350.
    PICE - 123, 448-451.

Chadwick, Sir Edwin, 1801-90, sanitary reformer.
    Iron - 36, 32-33.

Chadwick, William, 1810-70, ironworks representative.
    PIME - 1871, 15.

Challenor, Charles,   -1893, colliery owner.
    Iron - 41, 143*.

Chalmers, James, 18(19)-68, consulting engineer.
    Eng - 27, 19*.
    Mech Mag - 90, 17-18.

Chamberlain, Humphrey, 1846-96, locomotive engr in South America.
    PICE - 125, 398-399.

Chambers, Arthur Marshall, 18(43)-98, coal & iron master.
    Eng - 86, 234.
    Engng - 66, 295.
    JISI - 54, 326.

Chambers, J   C   , 18(32)-97, telegraph engineer.
    Elec - 39, 596-597.
    Elec Rev - 41, 312.

Chambers, John A   ,   -1892, telegraph engineer.
    Elec Rev - 31, 410*.

Chambers, W   Hoole, 18(40)-97, colliery owner.
    Eng - 84, 254.

Chance, William,   -1856, glass manufacturer.
    Eng - 1, 80*.

Chapman, Alfred Crawhall, 1859-96, consulting mining engineer.
    PIME - 1896, 255.

Chapman, Arthur, 1842-84, colliery & ironworks manager.
    TCMCIE - 13, 11.

Chapman, Charles,   -1891, hydraulic engineer.
    TMAE - 1, 199.

Chapman, Hedley,   -1897, ironmaster.
    Eng - 83, 425.

Chapman, John, 1826-91, gas works manager.
PIME - 1891, 191.

Chapman, John George, 1831-82, coal & ironworks manager.
JISI - 21, 661-662.

Chapman, Lt. William, 1822-53, Bombay Engineer.
PICE - 14, 167-171.

Charlton, Francis, 1816-81, county surveyor of Northumberland.
PICE - 66, 375-377.

Charlton, Thomas, 1818-72, mining engineer.
Engng - 14, 399.

Charlton, William Anthony, 1854-1900, company representative.
TIESS - 44, 339-340.

Charnock, James, 1851-99, engineer in Russia.
PIME - 1899, 265-266.

Checkley, Thomas , 1834-80, colliery owner.
PIME - 1881, 2-3.

Cheek, Phillip, 1849-91, cement manufacturer.
TSE - 1891, 216.

Cheeseman, Frank, 1848-79, ironmaster.
PICE - 60, 405-406.

Cheffins, Charles Frederick, 1807-60, railway engineer.
PICE - 21, 578-580.

Cherrie, James Macallum,   -1900, iron & steel merchant.
JISI - 57, 251.

Chesney, Lt.-Col. Charles Cornwallis, 1826-76, Royal Engineer(D).
Eng - 41, 206.

Childe, Rowland, 1826-86, consulting mining engineer.
PICE - 87, 417-418.

Chisholm, John, 1777-1856, water engineer.
PICE - 16, 121.

Chittenden, Francis Sheldon, 18(36)-90, railway engr overseas.
PICE - 100, 410-411.

Chrimes, Richard, 18(19)-97, brass manufacturer.
Eng - 83, 451.

Christie, William Buchan, 1847-94, Public Works engr in India.
PICE - 120, 343-344.

Chubb, Harry, 1816-87, railway manager, &c.
    PICE - 91, 457-459.

Church, Jabez, 1824-75, consulting gas & water engineer(B).
    Eng - 39, 372.
    PICE - 41, 211-212.
    TSE - 1896, 241.

Church, Jabez, Jun., 1845-96, gas & water works engineer(B).
    Eng - 81, 327.
    Engng - 61, 410.
    PICE - 125, 399-400.

Churchill, John Fleming, 1829-94, Public Works engr in Ceylon.
    PICE - 118, 443-445.

Churchward, Henry William, 1849-77, engineer in South America.
    PICE - 53, 286-287.

Churchward, William Henry, 1817-96, admiralty dock engineer.
    PICE - 127, 388-389.

Clapham, Henry,   -1883, shipbuilder & owner.
    JISI - 23, 667-669.

Clapham, Robert Calvert, 1823-81, chemical works manager(B).
    PIME - 1882, 2-3.

Clark, Christopher Fisher, 1831-98, mining engineer.
    PIME - 1898, 530.

Clark, Daniel, 1819-72, engr to Peruvian navy.
    Engng - 13, 97.

Clark, Daniel Kinnear, 1821-96, locomotive engineer(B).
    Elec - 36, 470*.
    Eng - 81, 118.
    Engng - 61, 193-194.
    PICE - 124, 409-413.
    PIME - 1896, 92-94.

Clark, Edwin, 1814-94, telegraph engineer(B,D).
    Elec Rev - 35, 508.
    Eng - 78, 367.
    Engng - 58, 555-556.
    PICE - 120, 344-345.

Clark, George, 1815-85, marine engineer.
    PIME - 1885, 160.

Clark, George Thomas, 1809-98, iron and coal master(B,D).
    Eng - 85, 116.
    Engng - 65, 177.
    JISI - 53, 313-314.

Clark, George W   , 18(19)-90, shipbuilder.
    Engng - 49, 676*.

Clark, James, 1836-76, company representative.
    PIME - 1877, 18.

Clark, John, 1831-67, surveyor in Hong Kong.
    PICE - 30, 431.

Clark, John,   -1885, nautical inventor.
    Eng - 60, 321.

Clark, Josiah Latimer, F.R.S., 1822-98, telegraph engineer(B,D).
    Elec - 42, 33.
    Elec Rev - 43, 663-664.
    Eng - 86, 451.
    Engng - 66, 589-590.
    JIEE - 28, 667-672.
    PICE - 137, 418-432.

Clark, Leslie, 1828-83, engineer in India.
    PICE - 77, 350-352.

Clark, Octavus Deacon, 1833-94, engineer in India.
    PICE - 118, 456-459.

Clark, Thomas, 1804-57, water engineer.
    PICE - 17, 96-97.

Clark, Thomas, 1812-73, marine engine builder.
    PICE - 1874, 17.

Clark, William, 1821-80, civil engineer & inventor(B,D).
    PICE - 63, 308-310.
    PIME - 1881, 3.

Clark, William J   , 1847-87, engineering manufacturer.
    JISI - 31, 209.
    NECIES - 3, xxxv.
    TIESS - 31, 236.

Clark, William Tierney, F.R.S., 1783-1852, civil engineer(B,D).
    CEAJ - 15, 399.
    PICE - 12, 153-157.

Clarke, George Howard, 1849-92, locomotive superintendent in Brazi
    PICE - 112, 365-366.

Clarke, Hyde, 1815-95, general engineer, linguist(B).
    Elec Rev - 36, 207.
    Eng- 179,217+port.
    Engng - 59, 318.

Clarke, I   ,   -1892, telegraph engineer.
   Elec Rev - 30, 45*.

Clarke, John, 1825-90, iron founder.
   Engng - 49, 197.
   PIME - 1890, 171-172.

Clarke, John Charles James, 18(46)-89, telegraph engr in Egypt.
   Elec - 23, 326.

Clarke, John Howard, 1849-92, railway engineer.
   PICE - 112, 365-366.

Clarke, Seymour, 1814-76, railway manager (B).
   PICE - 44, 225-227.

Clarke, Thomas Charles, 1835-93, railway engineer in Brazil.
   PICE - 113, 333-334.

Clarke, William, 1831-90, engineering manufacturer.
   NECIES - 6, xlii.

Clarke, William Welham,   -1877, Public Works engr in India.
   PICE - 52, 270.

Clarkson, Clement,   -1900, electrical engineer.
   Elec Rev - 47, 225*.

Clay, William, 1823-81, ironmaster(B).
   PIME - 1882, 3-5.

Clayton, Nathaniel, 1811-90, agricultural machinery manufr(B).
   Eng - 70, 527; 71, 18.
   Engng - 51, 17-19.
   PIME - 1890, 554-555.

Clayton, Richard Smith, 1838-76, civil engineer.
   PICE - 47, 297-298.

Clegg, Samuel, 1781-1861, gas engineer(B,D).
   PICE - 21, 552-554.

Clegg, Samuel, 1814-56, Prof. of   Engng & Architecture, Putney
   College(B,D).
   Eng - 2, 410.
   PICE - 16, 121-124.

Clegram, William, 1784-1863, canal engineer.
   PICE - 23, 485-486.

Clegram, William Brown, 1809-89, canal engineer.
   PICE - 98, 388-390.

Cleminshaw, Thomas Stanley, 1851-99, gas engineer.
PICE - 137, 439.

Cleminson, James Lyons, 1810-96, consulting civil engineer.
JISI - 50 , 257-258.

Cleminson, James Lyons, 1840-96, consulting railway engr.
Eng - 82, 530*, 545.
Engng - 62, 673.
PICE - 127, 379-380.
PIME - 1896, 255-256.

Cleminson, John,   -1878, railway engineer.
Eng - 45, 226.

Clerk, Francis North, 1829-74, iron manufacturer.
PIME - 1875, 20.

Clerke, William John Bird, 1838-96, sanitary inspecting engr.
Eng - 81, 195.
Engng - 61, 266.
JISI - 49, 283.
PICE - 125, 400-401.

Cleworth, Charles, 1831-75, loco. superintendent in India.
PIME - 1876, 18.

Clift, John Edward, 1817-75, gas engineer.
PIME - 1876, 18-19.

Close, John Forbes, 1857-94, dredging engr in Egypt.
PICE - 117, 390-391.

Clutton, John, 1809-96, surveyor(B).
PICE - 125, 430-434.

Coates, George, 1838-80, draughtsman.
PICE - 61, 297.

Coates, Major,   -1870, ironmaster.
Eng - 30, 34*.

Cochrane, Alexander Brodie, 1813-63, ironmaster & manufr.
PICE - 23, 506.
PIME - 1864, 13-14.

Cochrane, Charles, 1835-98, ironfounder(B).
Eng - 85, 464, 491.
Engng - 65, 636-637.
JISI - 53, 314-315.
PIME - 1898, 309-311.

Cochrane, John, 1823-91, civil engineering contractor.
PICE - 109, 398-399.

Cock, William Henry, 1833-86, surveyor in South America.
PICE - 87, 447-448.

Cockayne, Octavius, 18(26)-61, railway engineer.
PICE - 21, 581.

Coddington, John George Thornton, 1841-91, engr in India.
PICE - 104, 291-293.

Coddington, Capt. Joshua William, 1802-53, Royal Engineer;
Inspector of Railways.
PICE - 14, 165-167.

Cogar, William Abram, 1839-96, Secy, Gas Meter Company.
TSE - 1896, 242.

Coghlan, John, 1824-90, entrepreneur in River Plate area.
PICE - 103, 363-365.

Coke, Richard George, 1813-89, consulting mining engineer.
PICE - 96, 317-318.
TCMCIE - 17, 266-267.

Colburn, Zerah, 1832-70, American editor of "Engineer"(B).
Eng - 29, 317.
Engng - 9, 361.
Mech Mag - 92, 371.
PICE - 31, 212-217.
PIME - 1871, 15-16.

Colby, Maj.-Gen. Thomas Frederick, 1784-1852, Royal Engineer;
Director, Ordnance Survey(B,D).
CEAJ - 15, 399.
PICE - 12, 132-137.

Cole, George,   -1881, municipal surveyor.
PIMCE - 8, 227-228.

Cole, John William, 1840-91, mechanical engineer.
PIME - 1891, 290-291.

Cole, William Richard, 1829-82, mining engineer.
PICE - 73, 355-356.

Cole-Baker, George Hannyngton, 1866-94, railway engineer.
PICE - 119, 405-406.

Coleman, Joseph James, 1838-88, refrigeration pioneer(B).
Eng - 67, 4.
Engng - 46, 604*.
Iron - 32, 569.
TIESS - 32, 321.

Coles, Capt. Cowper Phipps, R.N., 1819-70, naval engr(B,D).
Engng - 10, 210.

Collett, Alfred, 1854-95, railway engr in South America.
PICE - 122, 364-365.

Colley, Thomas Long, 1840-68, water engineer.
PICE - 30, 466-467.

Collinge, Charles, 1792-1842, ---
PICE - 2, 13*.

Collingwood, Capt. Carlton Thomas,   -1860, ---
PICE - 22, 636*.

Collins, James, 1817-75, railway engineer.
PICE - 42, 257-258.

Collins, John,, 1832-88, metallurgist & chemist.
Engng - 46, 527.

Collins, John,   -1892, ironworks manager.
Iron - 39, 298.

Collins, William Whitaker, 1817-79, consulting engineer.
Engng - 27, 241.
PICE - 58, 340-341.

Colomb, Vice-Adm.Philip Howard, R.N., 1831-99, telegraph engr(B,D)
Elec Rev - 45, 646*.
Eng - 88, 403+port.
Engng - 68, 510.
TINA - 42, 295.

Colquhon, James, 1833-93, coal & ironworks manager.
JISI - 45, 390.
PIME - 1893, 492.

Colquhon, Lt.-Col. James Nisbet, F.R.S., 1791-1853, Royal
Artillery(B).
PICE - 13, 149-156.

Colthurst, Joseph, 1812-82, railway engineer.
PICE - 73, 356-358.

Colville, David, 1813-98, iron & steel manufacturer.
Eng - 86, 442.
Engng - 66, 581, 714.

Comber, Arthur, 1851-1900, municipal engineer.
TSE - 1900, 268.

Comrie, Alexander, 1786-1855, surveyor.
PICE - 15, 96-97.

Conder, Francis Roubiliac, 1815-89, railway engineer.
    Eng - 68, 517.
    PICE - 100, 379-383.

Condie, John, 1795-1860, mechanical engineer.
    Eng - 10, 309*.

Connell, Charles,   -1884, shipbuilder.
    Engng - 37, 187*.
    TIESS - 27, 213-214.

Conner, Benjamin, 1814-76, locomotive superintendent.
    Engng - 21, 117*, 135-136.

Connolly, Thomas,   -1899, electrical manufacturer.
    Elec Rev - 45, 27.

Connor, Maj.Augustus Samuel William, 1844-88, Bombay Engineer.
    PICE - 95, 396.

Conquest, Alexander William, 1848-92, sanitary engineer.
    PICE - 110, 384.

Conway-Gordon, Brevet-Col.Lewis, 1838-95, Royal Engineer(B).
    PICE - 122, 404-407.

Coode, Sir John, 1816-92, harbour engineer(B,D).
    Elec - 28, 448*.
    Eng - 73, 197.
    Engng - 53, 298.
    Iron - 39, 209.
    PICE - 113, 334-343.

Cooke, Alfred Thomas, 1819- 95, civil engineer.
    PICE - 121, 332-333.

Cooke, George Caesar,   -1875, engineer in India.
    PICE - 44, 227-228.

Cooke, James Samuel, 18(30)-87, railway surveyor.
    PICE - 90, 419-420.

Cooke, Joseph, 18(40)-99, steel manufacturer.
    Eng - 87, 423.

Cooke, Joseph Green, 1833-81, railway engineer overseas.
    PICE - 66, 380-382.

Cooke, Sir William Fothergill, 1806-79, telegraph engineer(B).
    Eng - 48, 74.
    Engng - 28, 94-95.
    JSTE - 8, 361-397. By Latimer Clark.
    PICE - 58, 358-364.

Cooper, Edmund, 1818-93, Board of Works engineer.
    PICE - 114, 368-369.
    TSE - 1893, 237.

Cooper, Frederick, 1836-82, armament company manager.
    PIME - 1883, 15-16.

Cooper, James, 1817-62, manufacturer.
    PICE - 22, 624-625.

Cooper, Matthew, 18(46)-99, Post Office engineer.
    Elec Rev - 45, 441.
    JIEE - 29, 948-949.

Cooper, Richard Williams 18(27)-91, Parliamentary solicitor.
    TSE - 1891, 216.

Cooper, Samuel Thomas, 1831-71, ironmaster.
    PICE - 33, 251.
    PIME - 1872, 16.

Cooper, William, 1831-87, shipping engineer in Bombay.
    PICE - 89, 491-492.

Cope, James, 18(10)-80, mining engineer.
    Engng - 30, 240.

Cope, James, 1818-70, consulting mining engineer.
    PIME - 1871, 16.

Copley, John Singleton, Baron Lyndhurst, F.R.S., 1772-1863,
    amateur engineer(D).
    PICE - 23, 478-479.

Corlett, Henry Lee, 1826-83, tramway company chairman.
    PICE - 74, 293-294.

Corry, Alfred James, 1858-92, water engineer.
    PICE - 110, 385-386.

Corry, Edward, 1817-1900, ore merchant.
    PICE - 140, 285-286.

Cory, John,   -1891, shipowner.
    Iron - 38, 541.

Cory, William,   -1868, shipowner & iron merchant.
    Eng - 25, 148*.

Cotton, E   J   , 1845-99, railway manager.
    Eng - 87, 621*.

Coulson, William, 1816-94, mining engineer.
    PIME - 1894, 276.

Coultas, James, 1819-90, ironmaster.
    Eng - 70, 335.

Coulthard, William, 1796-1863, railway engineer.
    PICE - 23, 506-507.

Coulthard, William Robson,   -1866, railway engineer.
    PICE - 28, 618-619.

Coulthurst, Thomas, 1838-88, municipal engineer.
    PIMCE - 14, 399-400.
    TCMCIE - 17, 267-270.

Cousins, Edward, 1820-99, municipal engineer.
    PICE - 140, 268.

Cowan, William, 1823-98, locomotive superintendent.
    Eng - 85, 261.
    TIESS - 41, 378-379.
    TSE - 1898, 233.

Cowen, John Anthony, 1831-95, firebrick manufacturer.
    TSE - 1895, 270.

Cowley, John James, 1863-98, railway engineer.
    PICE - 134, 414.

Cowlinshaw, John, 1848-84, mining engineer.
    TCMCIE - 13, 11-12.

Cowper, Charles, 1821-60, industrial chemist.
    PICE - 21, 581-582.

Cowper, Ebenezer, 1804-80, printing innovator(B,D).
    Engng - 30, 257.

Cowper, Edward Alfred, 1819-93, consulting mechanical engineer(B).
    Elec - 31, 67.
    Elec Rev - 32, 627.
    Eng - 75, 429.
    Engng - 55, 712-713.
    Iron - 43, 429-430.
    JISI - 42, 172-173.
    PICE - 114, 369-372.
    PIME - 1893, 203-205.
    TINA - 34, 241-242.

Cox, Samuel Fitzhugh, 1847-87, railway engineer in India.
    PICE - 92, 384-387.

Coxhead, Frederick Carley, 1828-95, consulting engineer.
    PIME - 1895, 142.

Coxon, Samuel Bailey, 1834-87, mining engineer.
    PICE - 92, 387-388.

Coy, Joseph Prendergast, 1857-92, Public Works engr in India.
    PICE - 110, 386-387.

Crabtree, Henry, 18(37)-82, railway engineer in Argentina.
    PICE - 70, 434.

Crabtree, William, 1826-96, municipal engineer of Stockport.
    PICE - 125, 401-403.
    PIMCE - 22, 363.

Cracknell, Edward Charles, 1831-93, telegraph engr in Australia(B)
    Elec - 30, 319.
    Elec Rev - 32, 240-241.
    PICE - 113, 343-344.

Craggs, J   ,   -1880, municipal engineer.
    PIMCE - 7, 147*.

Craggs, William, 18(21)-95, engine driver.
    Eng - 80, 525.

Craig, James, 1844-97, Public Works engineer in India.
    PICE - 129, 367-368.

Crampton, Thomas Russell, 1816-88, telegraph engineer(B,D).
    Elec - 20, 711.
    Elec Rev - 22, 443.
    Eng - 65, 348.
    Engng - 45, 415.
    Iron - 31, 355.
    JIEE - 17, 421-423.
    JISI - 32, 210-213.
    PICE - 94, 295-298.
    PIME - 1888, 437-439.

Crampton, Willoughby, 18(45)-97, manufacturer.
    Eng - 84, 355.

Crane, George,   -1846, ironmaster.
    PICE - 6, 5*.

Craven, John, 18(35)-1900, tool manufacturer.
    Eng - 89, 621.
    JISI - 57, 252.
    TMAE - 10, 365-366.

Craven, John Chester, 1813-87, locomotive superintendent(B).
    Engng - 44, 71.
    Iron - 30, 89.
    PICE - 90, 420-423.

Craven, Joseph, 18(18)-1900, carriage & wagon builder.
JISI - 59, 316.

Craven, Thomas Edwin, 1821-92, consulting engineer.
PIME - 1892, 404.

Crawford, William, M.P., 1833-90, Secy, Miners' Association.
Iron - 36, 9.

Crawhall-Wilson, Thomas Wilson,   -1892, coal & ironmaster.
JISI - 42, 297-298.

Crawley, George Baden, 1833-79, railway contractor overseas(B).
Engng - 28, 434.

Crawshay, Edmund,   -1900, ironmaster.
Eng - 89, 604*.

Crawshay, Henry,18(02)-79, ironmaster.
Eng - 48, 403-404.
Engng - 28, 423.
Iron - 14, 681.
JISI - 15, 614-615.

Crawshay, Robert Thompson, 1817-79, ironmaster(B,D).
Eng - 47, 359.
Engng - 27, 416.
JISI - 14, 328.
Prac Mag - 1873, 81-84+port.

Crawshay, William, 17(85)-1867, ironmaster.
Mech Mag - 87, 96.

Creswick, James Frost,   -1895, admiralty engineer.
Engng - 57, 563.

Cresy, Edward, 1824-70, engineering writer.
Eng - 30, 409.

Crichton, Robert, 1847-95, railway engineer.
PICE - 121, 324-325.

Croker, Bland William, 1822-72, civil engineer.
Eng - 32, 140.

Croll, Alexander Angus, 1811-87, telegraph engineer(B).
Elec Rev - 20, 580*.
PICE - 90, 446-449.

Crommelin, Lt.-Gen. William Arden, 1823-86, Bengal Engineer(B).
Eng - 62, 366.

Crompton, George, 1823-97, ironmaster.
JISI - 53, 315*.

55

TFIME - 16, 125-126.

Crompton, William,   -1892, colliery owner.
    Iron - 40, 384*.

Crook, Charles Richard Ernest, 1863-92, civil engineer.
    PICE - 115, 398-399.

Crosland, William Mann, 1824-89, marine engineer.
    PICE - 99, 373-374.

Crosley, William, 1819-74, gas engineer.
    PICE - 41, 224-225.

Cross, James, 1829-94, civil engineer.
    PICE - 119, 390.
    PIME - 1894, 464.

Cross, Robert James, 1841-93, shipping engineer.
    PIME - 1893, 90.

Cross, William, 1843-99, gas engineer.
    PICE - 139, 376.

Crossland, James, 1831-97, admiralty engineer.
    TINA - 40, 293.

Crossley, Francis William, 18(40)-97, gas engineer & manufr.
    Elec Rev - 40, 463*, 499.
    Eng - 83, 351÷port.
    Engng - 63, 441.

Crossley, John, 18(39)-1900, ---
    Eng - 90, 393*.

Crossley, John Sydney, 1812-79, railway engineer(B).
    PICE - 58, 341-343.

Crossley, Louis John, 1842-91, telegraph engineer.
    Elec - 27, 523-524.
    Elec Rev - 29, 280.
    Iron - 38, 210.

Crowe, Edward, 1829-73, ironworks engineer.
    PIME - 1874, 17-18.

Crowe, William Milner, 1846-88, civil engineer.
    PICE - 93, 496-497.

Crowley, William Henry, 18(38)-98, ironmaster.
    Eng - 86, 167.
    Engng - 66, 204.

Crowther, Clement, 1835-96, ironmaster.
    JISI - 49, 284-285*

Crowther, Thomas Burnside, 1866-1900, rly engineer in Latin
    America.
    PICE - 143, 329-330.

Cruse, Alfred Brace, 1843-84, railway engineer.
    PICE - 82, 383-385.

Cruse, Thomas, 1809-88, municipal engineer.
    PIMCE - 14, 205*; 15, 205-206.

Cryer, Thomas, 1849-92, mechanical engineering lecturer.
    TMAE - 2, 277-278.

Cubitt, Benjamin, 1795-1848, railway superintendent.
    PICE - 8, 10.

Cubitt, Joseph, 1811-72, railway engineer(B,D).
    PICE - 39, 248-251.

Cubitt,Thomas, 1788-1855, builder(B,D).
    PICE - 16, 158-162.

Cubitt, Sir William, F.R.S., 1785-1861, civil engineer(B,D).
    CEAJ - 24, 343, 345.
    Eng - 12, 230.
    PICE - 21, 554-558.

Cubitt, William, 1791-1863, builder(B,D).
    PICE - 23, 507-508.

Cudworth, James Ianson, 1817-99, locomotive superintendent.
    Eng - 88, 433*, 479.

Cuming, James Hening, 1863-98, railway engr in India.
    PICE - 137, 440.

Cumming, W   S   ,   -18(92), draughtsman.
    TIESS - 36, 317-318.

Cundy, John, 1828-92, railway engineer.
    PICE - 109, 415-416.

Cunliff, Richard Stedman, 1804-79, ship owner.
    Engng - 27, 56.

Cunningham, David, 1838-96, harbour engineer.
    PICE - 126, 393-394.
    TIESS - 39, 269.

Cunningham, George M   ,   -1897, railway contractor.
    Engng - 63, 440.

57

Cunningham,Henry Duncan Preston, 1815-75, inventor(B).
    Eng - 39, 85.
    Engng - 19, 83.
    TINA - 16, 273-275.

Curll, William Henry Richards, 1828-65, rly engr in India.
    PICE - 25, 526-528.

Curran, Lt. Isaac, 1818-76, armaments engineer
    PICE - 28, 619-621.

Currey, Charles, 1833-78, railway engineer in India.
    PICE - 55, 327-329.

Currey, Henry, 1820-1900, architect & surveyor.
    PICE - 143, 339-341.

Currie, James, 1822-1900, ship owner.
    Eng - 89, 227, 235.

Curtis, Charles Berwick, 1795-1877, gunpowder manufacturer(B).
    PICE - 49, 266.

Curtis, John George Cockburn, 1817-79, marine engineer.
    PICE - 60, 393-395.

Cuthbert, Thomas, 18(01)-92, railway coachbuilder.
    Iron - 39, 185*.

Daft,Thomas Barnabas, 1816-78, works manager(B).
    PICE - 55, 329-330.

Daglish, Robert, 1779-1865, colliery engineer(B).
    Eng - 21, 31*.
    PICE - 26, 561-563.

Dale, Edward J   ,   -1900, electrician.
    Elec Rev - 46, 710*.

Dale, Thomas, 1819-75, municipal engineer.
    PICE - 41, 212-214.

Dalgleish, Robert, 1809-83, engineering manufacturer.
    JISI - 25, 555.
    PICE - 74, 283-285.

Dalrymple, Grant Samuel, 1816-51, civil engineer.
    PICE - 11, 99.

Dangerfield, Henry, 1844-87, Public Works engineer in India.
    PICE - 90, 423-428.

Daniel, B   ,   -1899, colliery manager.
    Eng - 87, 634*.

Daniels, Thomas, 1841-1900, works manager.
    Eng - 89, 293.
    PIME - 1900, 328-329.
    TIESS - 43, 355-356.
    TMAE - 10, 361-364.

Danks,   Thomas Albert, 18(63)-1900,   ---
    Eng -   89, 541*.

Darby, Abraham, 1803-78, ironmaster.
    Engng - 26, 477*.
    JISI - 13, 612-615.
    PIME - 1879, 9.

Darby, Charles Edward, 1822-84, ironmaster.
    PIME - 1884, 400.

Darley, George Johnstone, 1822-98, land agent.
    PICE - 132, 376-377.

Darling, John, 1835-97, marine superintendent.
    TIESS - 40, 249-250.

Darwent,   , 18(25)-73, telegraph engr in Australia.
    Engng - 15, 6*.

Dash, Thomas Alexis, 1827-97, land surveyor.
    PICE - 128, 362-363.

Davenport, Edward Gershom, M.P., 1838-74, railway engr(B).
    PICE - 41, 225-226.

Davidson, Alfred, 1845-1900, Public Works engineer in India.
    Eng - 90, 191*.
    PICE - 143, 311-312.

Davidson, James, 1819-89, armaments engineer.
    PIME - 1889, 332-333.

Davidson, James, 1836-86, civil engineer overseas.
    PIME - 1887, 146.

Davie, William,   -1870, railway engineer in India.
    Engng - 9, 166*.

Davies, Charles, 1818-79, railway engineer in India.
    PICE - 68, 314-315.

Davies, Charles Lennox, 1822-82, engineer in India.
    PICE - 71, 398-399.

Davies, David, 18(18)-90, railway contractor.
    Iron - 36, 80.

Davies, David Christopher, 18(27)-85, mining engineer(B).
    Iron - 27, 289.

Davies, David George, 1859-98, consulting railway engineer.
    PICE - 135, 364-365.

Davies, Edward,   -1898, railway director.
    Eng - 85, 23.

Davies, John, 18(34)-99, mining engineer.
    Eng - 87, 504.

Davies, Lee,   -1900, surveyor & mining engineer.
    Eng - 89, 472*.

Davies, Rhys, 18(45)-99, municipal engineer.
    PIMCE - 25, 479*.

Davies, Thomas George, 18(55)-99, hydraulic engineer.
    Eng - 88, 180.

Davies, W   H   ,   -1888, telegraph engineer.
    Elec - 21, 332.

Davies, William, 1836-97, iron works manager.
    JISI - 52, 254.

D'Avigdor, Elim Henry, 1841-95, civil engineer.
    PICE - 121, 340-341.

Davis, E   ,   -1900, manufacturer.
    Elec Rev - 47, 795.

Davis, Francis Gordon, 1843-69, mining engineer.
    PICE - 30, 467-468.

Davis, Frederick, 1843-1900, engineering manufacturer.
    PICE - 142, 386.

Davis, John Henry, 1837-96, consulting engineer.
    PICE - 126, 394-395.

Davis, Joseph, 1839-93, railway engineer.
    PIME - 1893, 492-493.
    TMAE - 3, 284-285.

Davis, Thomas, 1810-77, ironmaster.
JISI - 11, 539-540 *.

Davison, Robert, 1804-86, brewery engineer(B).
PICE - 84, 442-444.

Davison, Samuel, 1837-83, ironworks manager.
JISI - 23, 671.

Davison, Samuel Dobson, 1821-83, manufacturer.
Engng - 36, 453.

Davy, Abraham, 1833-96, iron & colliery director.
Eng - 81, 111.
JISI - 49, 285*.

Davy, Walter Scott, 1832-87, ironmaster.
PIME - 1887, 274.

Dawes, W   H   ,   -1878, ironmaster.
JISI - 12, 295-296.

Dawson, Bernard, 1851-1900, consulting engineer.
JISI - 57, 252.
PIME - 1900, 329-330.

Dawson, Francis,   -1879, engineer in Jamaica.
PICE - 59, 313-314.

Dawson, Lt.-Col. Robert Kearsley, 1798-1861, Royal Engineer(B,D).
PICE - 21, 582-584.

Dawson, T   Galloway, 18(29)-99, manufacturer.
Eng - 87, 413*.

Day, Edward Hugh, 1868-96, railway engineer.
PICE - 126, 400-401.

Day, Richard,   -1900, colliery owner.
JISI - 59, 316-317.

Deakin, James, 18(22)-1900, steel manufacturer.
Eng - 89, 105.

Dean, John William, 1844-96, railway contractor.
JISI - 49, 285.

Deane, John Horridge, 1828-61, railway engineer overseas.
PIME - 1862, 14.

Deas, James, 1827-99, marine engineer.
Eng - 89, 10+port.
Engng - 69, 23-24+port.

PIME - 1900, 330-331.
TIESS - 43, 356-358.

Debenham, John,   -1866, civil engineer.
    Eng - 21, 72*.

De Bergue, Charles Louis Aimé, 1807-73, civil engng contractor(B).
    PICE - 38, 309.
    PIME - 1874, 18-19.

Dees, James, 1815-75, rly director & colliery owner.
    PICE - 43, 297-298.
    PIME - 1876, 19-20.

Delaney, Edward Magdalen Joseph, 1836-66, surveyor.
    Eng - 24, 28.
    PICE - 27, 599-600.

Delany, Joseph Francis, 1833-81, shipbuilder.
    PICE - 65, 375.

De Leonval, Thomas Fletcher Chapee, 1824-95, mechanical engr &
    cotton spinner.
    PICE - 120, 342-343.

Dempsey, George Drysdale,   -1859, railway engineer.
    Eng - 8, 465*.

Dempsey, William, 1817-93, railway consulting engineer.
    PICE - 115, 385-386.

Denham, Adm. Sir Henry Mangles, F.R.S., 1800-87, hydrographer(B).
    PICE - 91, 460-462.

Denison, H   D   ,   -1889, manufacturer.
    Eng - 68, 100*.

Denison, Lt.-Gen. Sir William Thomas, 1804-71, Royal Engr(B,D).
    PICE - 33, 251-259.

Dennis, William Frederick, 1844-90, wire netting manufacturer.
    JISI - 36, 184-185.
    PIME - 1889, 748.

Denny, Peter, 1821-98, shipbuilder (B).
    Eng - 80, 187*, 192*.
    Engng - 60, 243, 276-277.
    PICE - 123, 423-432.
    TIESS - 38, 332-333. Port.- frontis. v.30.
    TINA - 37, 395.

Denny, Robert, 1843-88, railway engineer.
    PICE - 95, 369-371.

Denny, William, 1847-87, shipbuilder & marine engr(B).
    Eng - 63, 237.
    Engng - 43, 281-283.
    Iron - 29, 254.
    JISI - 31, 207-208.
    PICE - 89, 457-466.
    PIME - 1887, 274-276.
    TIESS - 30, 258-286+port, 310. By John Ward.
    TINA - 28, 455-458+port.

Denroche, Charles,   -1855, sanitary engineer.
    PICE - 15, 103.

Dent, Adm. Charles Bayley Calmady, 1832-94, marine superintendent,
    London & North-Western Railway(B).
    Eng - 77, 245*.

Dent, Edward John, 1790-1853, chronometer maker(B,D).
    PICE - 13, 156-161.

Dent, John Dent, 1826-94, railway chairman(B).
    Eng - 78, 571.

Dent, Montagu Charles,   -1900, consulting electrical engr.
    Elec Rev - 47, 23.

Denton, John Bailey, 1814-93, sanitary engineer(B).
    Eng - 76, 489.
    Engng - 56, 665.
    PICE - 115, 386-393.

Depree, Charles Lambert, 1845-93, consulting engr in Australia.
    PICE - 115, 389-391.

De Sauty, Charles Victor, 1830-93, telegraph superintendent.
    Elec - 30, 685.
    Elec Rev - 32, 431*.

Despard, Fitzherbert Ruxton, 1841-95, civil engineer overseas.
    PICE - 122, 396-397.

Despard, Richard Carden, 1831-63, civil engineer.
    PICE - 23, 486-487.

Deville, James,   -1846, lighting manufacturer.
    PICE - 6, 5*.

Devine, William Henry, 1844-99, mechanical engr in Japan.
    PIME - 1899, 613.

Devonshire, 7th Duke of, 1808-91, industrialist(B,D).
    JISI - 39, 120-127.
    PICE - 107, 393-396.
    TIME - 3, 1012-1013.

Dewrance, John, 1804-61, railway engineer.
    Eng - 12, 36*.

Dick, Charles, 1838-88, locomotive works manager.
    PICE - 94, 321-322.

Dickeson, Sir Richard, 1823-1900, military contractor(B).
    Eng - 90, 392*.

Dickinson, George, 18(21)-1900, surveyor.
    Eng - 89, 576.

Dickinson, Richard Elihu, 1849-95, ironmaster & manufacturer.
    JISI - 47, 261*.
    NECIES - 12, 246.
    PICE - 123, 432-433.
    PIME - 1896, 309-310.
    TFIME - 12, 124-125.

Dickson, George Manners, 1844-95, mechanical engineer.
    PIME - 1896, 598.

Dickson, George Workman, 1847-1900, colonial civil engineer.
    Eng - 90, 4.
    Engng - 70, 70.
    PICE - 141, 342.

Dieckstahl, George Charles, 1840-80, machinery manufacturer.
    Iron - 15, 450.
    JISI - 17, 689-690.

Dinnen, John, 1808-66, admiralty machinery inspector.
    Eng - 21, 31*.
    PICE - 26, 563-565.

Ditchburn, Thomas Joseph, 1801-70, shipbuilder.
    Eng - 29, 265; 87, 82-83+port.
    TINA - 11, 233-246; 12, 306.

Dixon, Charles William, 1841-71, theoretical engineer.
    PICE - 36, 288.

Dixon, Edward, 1809-77, railway contractor.
    PICE - 54, 280-281.

Dixon, John, 1835-91, contractor overseas(B).
    Eng - 71, 106.
    Engng - 51, 168-169.
    Iron - 39, 537.
    PICE - 104, 309-311.

Dixon, William Smith, 1824-80, ironmaster.
    Engng - 29, 501.
    Iron - 15, 450.

Dobson, Benjamin, 1823-74, machinery manufacturer.
PIME - 1875, 20-21.

Dobson, Sir Benjamin Alfred, 1847-98, cotton machinery manufr(B).
Eng - 85, 238+port.
Engng - 65, 311.
PIME - 1898, 136-137.
TMAE - 8, 217-218.

Dobson, George Clarisse, 1801-74, dock engineer(B).
PICE - 41, 214.

Dobson, Samuel, 1826-70, mining engineer.
PICE - 31, 217-218.

Dockray, Robert Benson, 1811-71, railway engineer(B).
PICE - 33, 213-215.

Dodds, Isaac, 1801-82, locomotive builder(B).
PICE - 75, 308-314.

Dodds, Thomas Weatherburn, 1826-99, rly plant manufacturer.
Eng - 88, 266.
Engng - 68, 328.
PICE - 139. 351-352.

Dodson, Arthur John, 1818-76, engineer in India.
PICE - 45, 243-244.

Dodson, Edward, 1841-83, steel manufacturer.
PIME - 1884, 62.

Doherty, William James, 1834-98, contractor.
PICE - 135, 365-366.

Donaghue, John, 1866-93, waterworks engineer.
PICE - 116, 376-377.

Donaldson, George Gardner, 1801-59, surveyor.
PICE - 19, 186-187.

Donaldson, John, 1841-99, marine engineer(B).
Eng - 88, 354.
Engng - 68, 427*, 464-466+port.
PICE - 140, 270-273.
PIME - 1899, 613-614.

Donaldson, Robert, 18(27)-85, iron merchant.
Eng - 40, 503*.
Iron - 26, 461.

Donaldson, William, 18(38)-1900, hydraulic engineer.
Eng - 89, 520.
Engng - 69, 684.

Doncaster, Charles,   -1885, steel works manager.
    Iron - 25, 16.

Donkin, Bryan, 1809-93, paper manufacturer(B).
    Elec - 32, 156*.
    Eng - 76, 525.
    Engng - 56, 696.
    PICE - 115, 391-393.

Donkin, John, 1802-54, paper maker.
    PICE - 14, 130-131.

Dorman, Robert Page, 18(56)-98, manufacturer.
    Eng - 86, 360.
    JISI - 54, 326.

Dorning, Elias, 1819-96, railway engineer.
    PICE - 126, 395-396.

Dossor, Arthur Loft, 1843-81, company representative.
    PIME - 1882, 5.

Doubleday, Thomas, 1790-1870, coal trade secretary(B,D).
    Engng - 10, 463*.

Douglas,   ,   -1892, marine engine manufacturer.
    Eng - 73, 197*.

Douglas, Gen. Sir Howard, Bart., F.R.S., 1776-1861, military
    engineer(B).
    Eng - 12, 296*.

Douglas, John, 1836-92, gas engineer.
    PICE - 112, 366-367.

Douglass, Sir James Nicholas, F.R.S., 1826-98, lighthouse engin-
    eer(B,D).
    Elec - 41, 275.
    Elec Rev - 41, 770.
    Eng - 85, 601.
    Engng - 65, 801.
    JIEE - 28, 672.
    PICE - 134, 403-405.
    PIME - 1898, 531-532.
    TSE - 1898, 238.

Douglass, William,   -1898,   ironmaster.
    JISI - 53, 315.

Doulton, Sir Henry, 1820-97, sanitary ware manufr(B,D).
    Elec - 40, 111*.
    Eng - 84, 490.
    PICE - 131, 391-394.

Dowden, William George, 1849-96, iron works manager.
JISI - 49, 258*.

Downes, Charles Campbell, 1837-87, rly engineer overseas.
PICE - 92, 394-396.

Downie, John,   - 1875, dynamite manufacturer.
Engng - 19, 346.

Downing, Samuel, 1811-82, Prof. of Civil Engng, Trinity College,
Dublin(B).
PICE - 72, 310-311.
PIME - 1883, 16-18.

Downing, Samuel Forbes, 1844-91, lecturer in India.
PICE - 109, 416-418.

Dowrie, Thomas, 18(30)-90, engineer in India.
TSE - 1890, 211.

Doyne, William Thomas, 1823-77, consulting engr in Australia(B).
PICE - 52, 270-273.

Dransfield, Major William,   -1898, railway contractor.
Eng - 85, 361.

Draper, Capt.R   ,   -1888, cable ship captain.
Elec Rev - 23, 629.

Dredge, James, 1848-76, railway engineer.
PICE - 49, 267-268.

Dresing, Peter Christian,1852-98, telegraph engineer.
JIEE - 28, 672-673.

Drew, Napoleon Edward, 1864-97, railway engineer overseas.
PICE - 128, 360.

Driver, Charles Henry, 1832-1900, design engineer.
PICE - 143, 341-342.

Druce, E   ,   -1898, civil engineer.
Eng - 85, 285.

Drummond, Richard Oliver Gardner, 1862-98, engr in South Africa.
Elec Rev - 43, 168.
JIEE - 28, 673.
PICE - 134, 414-415.

Drummond, Capt.Thomas, F.R.S., 1797-1840, Royal Engineer(D).
CEAJ - 3, 164.
PIME - 1898, 533.

Drysdale, Alexander, 1817-83, civil engineer.
PICE - 75, 299-301.

Dübs, Henry, 1816-76, locomotive manufacturer.
    Engng - 21, 348*, 366.

Dübs, Henry John Sillars, 1848-93, locomotive manufacturer.
    PIME - 1893, 90-91.

Dudgeon, John, 1816-82, shipbuilder.
    Eng - 87, 83+port.

Dudgeon, William,   -1875, shipbuilder.
    Eng - 39, 255.

Dudley, 1st Earl of, 1817-85, ironmaster.
    JISI - 27, 536-537.

Duncan, Robert, 1827-89, shipbuilder.
    Eng - 68, 28.
    Engng - 48, 54.
    Iron - 34, 54.
    TIESS - 32, 321.

Duncan, Thomas, 1804-68, municipal engineer.
    Eng - 26, 440.
    PICE - 30, 431-434.

Duncan, Thomas, 1855-92, road engineer.
    PICE - 110, 387-388.

Dundas, James, 17(92)-1881, foundryman.
    Iron - 17, 172.

Dundas, Robert, 1838-97, railway engineer.
    Engng - 64, 12.
    PICE - 130, 312-313.
    TIESS - 40, 250-251+port frontis.

Dunkerley, Charles Chorlton,   -1898, iron trade representative.
    JISI - 54, 327.

Dunlop, Alexander Milne, 1841-97, surveyor.
    JISI - 51, 310.
    PICE - 127, 395-396.

Dunlop, Herbert Henry Grahame,1861-1900, rly engr in India.
    PICE - 143, 330.

Dunlop, James,   -1893, ironmaster.
    Iron - 41, 88.

Dunlop, John Macmillan, 1818-78, engineering manufacturer.
    PIME - 1879, 9.

Dunlop, Capt. Samuel John, 1845-78, engineer in India.
    PICE - 54, 283-284.

Dunn, James Benjamin, 1827-74, engineering agent.
    PICE - 40, 260-261.

Dunn, John George,   -1890, submarine telegraph engineer.
    Elec - 24, 573-574.

Dunn, Thomas, 1813-71, ironworks engineer.
    Eng - 31, 24*; 32, 435*.
    PICE - 36, 288.

Dunn, Thomas Edward, 1834-78, railway engineer in India.
    PICE - 53, 282-283.
    PIME - 1879, 10.

Dunning, John, 1827-85, ironmaster.
    Engng - 39, 265.
    PICE - 81, 342-343.

Dunscombe, Nicholas, 1867-96, municipal engineer.
    PICE - 128, 361.
    PIMCE - 23, 478-479.

Du Port, William Janvrin, 1834-91, dock engineer in Egypt(B).
    Iron - 37, 572.
    PICE - 106, 319-321.

Durham, 2nd Earl of, 1828-79, colliery owner.
    JISI - 15, 613.

Du Sautoy, C   S   , 18(59)-99, marine superintendent.
    Elec Rev - 41, 410.
    Engng - 68, 495.

Du Sautoy, John Burn Anstis, 18(49)-89, railway engineer.
    PICE - 96, 350-351.

Dutch, Ernest, 1866-99, draughtsman.
    PIME - 1899, 471.

Duthie, Alexander,   -1897, shipbuilder.
    Engng - 63, 853.

Dutton, Francis Stacker, 1818-77, Australian agent in London(B).
    PICE - 49, 268-270.

Dyer, Col. Henry Clement Swinnerton, 1834-98, Royal Artillery(B).
    Elec - 40, 712*.
    Eng - 85, 279+port.
    Engng - 65, 396.
    JISI - 53, 315-316.
    PICE - 133, 394-395.

Dykes, David Stewart, 1830-57, mechanical engr & shipbuilder.
    PICE - 17, 100.

Dymon, John,   -1868, ---
    Eng - 26, 180*.

Dyson, George, 1817-91, civil engr & steel manufacturer.
    JISI - 40, 128-129.
    PICE - 106, 347-348.

Eagles, Thomas Henry, 1845-92, engineering lecturer.
    PICE - 110, 392-393.

Earl, Henry Harrison, 18(10)-99, iron merchant.
    Engng - 67, 749.

Earle , C   W   , 18(29)-97, telegraph engineer.
    Elec - 39, 209.
    Elec Rev - 40, 822.

Eassie, Peter Boyd, 1835-75, railway contractor(B).
    PIME - 1876, 20.

Eassie, William, 1832-88, civil & sanitary engineer(B).
    Iron - 32, 199.
    TSE - 1888, 253.

East, Frederick, 1819-75, railway engineer in India.
    PICE - 43, 299-300.

Easton, Alexander, 1787-1854, canal engineer.
    PICE - 14, 131-133.

Easton, James, 1796-1871, civil engineer.
    Eng - 32, 296, 306-307.
    Engng - 12, 308.

Easton, James, 1830-88, pump manufacturer.
    PICE - 95, 370-371.

Eastwood, James, 1808-74, machine tool manufacturer.
    PIME - 1875, 21-22.

Eaves, William,   -1900, mechanical engineer.
    Engng - 70, 500*.

Eborall, Cornelius Willes, 1820-74, railway company manager(B).
    Eng - 36, 420-421.
    Engng - 16, 512.
    PICE - 39, 287-289.

Eckersley, William Alfred, 1856-95, railway engineer overseas.
    PICE - 122, 366-367.

Eddison, Robert William, 1835-1900, steam plough manufacturer.
    Eng - 89, 536, 541*.
    Engng - 69, 685.
    PIME - 1900, 624-625.

Eddy, Edward Miller Gard, 1851-97, rly engineer in Australia(B).
    Eng - 84, 131*, 161+port.
    Engng - 63, 860.
    PICE - 129, 400-403.

Eden, Hon. Francis Fleetwood, 1865-98, rly engineer in S.America.
    PIME - 1898, 311-312.

Edge, John Harris, 18(32)-1900, manufacturer.
    JISI - 58, 388.

Edgeworth, David Reid, 1842-71, railway engineer.
    PICE - 36, 264-265.

Edlin, Herbert William, 1856-97, mechanical engr overseas.
    PIME - 1898, 702.

Edwards, Henry Hinde, 1800-61, engng manufr in France.
    PICE - 22, 625-626.

Edwards, Osborne Cadwallader, 1822-76, consulting rly engr.
    PICE - 48, 266.

Eidsforth, Richard Pentney, 1844-99, electrical manufacturer.
    JIEE - 28, 673-674.

Elder, John, 1824-69, shipbuilder & marine engineer(B,D).
    Eng - 28, 217.

Eldridge, James, 1814-84, gas engineer & surveyor.
    PICE - 78, 441-442.

Eley, William, 1821-81, ammunition manufacturer(B).
    Iron - 18, 524.

Ellacott, John Hosking, 1811-94, draughtsman.
    PICE - 116, 353-354.

Elliot, Edmund Colville, 1857-93, Public Works engr in India.
    PICE - 116, 377-379.

Elliot, Frederick Henry, 1820-73, instrument maker.
    Engng - 15, 62*.

Elliot, Sir George, Bart., 1815-93, cable manufacturer(B).
    Elec - 32, 219-220.

Eng - 76, 617.
Engng - 56, 791.
JISI - 45, 390-391.
PICE - 116, 355-357.
Prac Mag - 4, 161-168+port.

Elliot, Liddle, 1807-69, civil engineer.
PICE - 31, 239.

Elliot, Ralph, 18(41)-75, telegraph engineer.
PICE - 41, 226.

Elliot, William, 1827-92, railway engineer in Brazil.
PICE - 111, 371-373.

Elliot, William, 1856-93, railway engineer in Brazil.
PICE - 116, 379-380.

Elliott, G   W   ,   -1898, mining engineer.
Eng - 86, 70.
Engng - 66, 78.

Elliott, Thomas Graham, 1847-99, works manager.
PIME - 1899, 127-128.

Ellis, Edward, 18(28)-97, surveyor.
TSE - 1897, 202.

Ellis, George,   -1857, ---
PICE - 17, 107*.

Ellis, J   J   , 18(61)-1900, admiralty engineer.
Engng - 69, 123.

Ellis, J   W   , 18(54)-91, ironmaster.
Iron - 38, 493.

Ellis, Thomas, 1818-84, ironmaster.
Engng - 38, 116.
Iron - 24, 119.
JISI - 25, 556-557.

Ellis, Thomas,   -1899, colliery manager.
Eng - 88, 102.

Ellis, Thomas Charles, 1847-86, water engineer.
PICE - 88, 443.

Ellis, Thomas Leonard,   -1897, iron works manager.
JISI - 54, 327.

Elsee, Thomas, 18(38)-1900, shipbuilder & marine engineer.
TIESS - 40, 340-341.

Elthringham, Joseph T   , 18(45)-97, manufacturer.
     Eng - 83, 652*.

Elwell, Paul Bedford, 1853-99, electrical engr in Australia.
     Elec Rev - 45, 646.
     Eng - 88, 432.
     PICE - 139, 352-353.

Elwell, Thomas, 1812-80, machinery manufacturer.
     PIME - 1881, 4.

Embleton, Thomas William, 1809-93, mining engineer.
     Eng - 76, 495.
     TIME - 7, 222-227+port, 494. By T.W.H. Mitchell.

England, Henry,   -1866, ---
     Eng - 21, 72*.

England, John, 1813-82, railway engineer overseas.
     PICE - 70, 415-417.

England, John, 1822-77, railway engineer overseas.
     PICE - 53, 283-284.

English, Thomas Alfred,   -1889, gas & water engineer.
     PICE - 97, 400-401.

Enys, John Samuel, 1796-1872, amateur engineer.
     PICE - 36, 290-293.

Epstein, Ludwig, 1853-99, electrician.
     Elec Rev - 45, 244.
     Eng - 88, 204*.
     Engng - 68, 205.
     JIEE - 29, 950-951.

Eraut, George, 1861-94, gas & water works manager.
     PICE - 117, 391-392.
     TSE - 1894, 253.

Erichsen, Hermann Gustav, 1828-91, telegraph engineer.
     Elec - 26, 163.
     JIEE - 20, 1.

Ericsson, Capt. John, 1803-89, inventor & engineer.
     Eng - 67, 229.
     Engng - 47, 258-259; 69, 354*.
     Iron - 33, 229.

Errington, John Edward, 1806-62, railway engineer(B,D).
     Eng - 14, 22*.
     PICE - 22, 626-629.

Erskine, George Holland, 1858-85, engineer in Goa.
    PICE - 88, 443.

Eskholme, George, 18(19)-98, mechanical engr & manufacturer.
    Engng - 65, 112.

Estridge, Lt.-Col. Joseph, 1811-90, Royal Engineer.
    PICE - 106, 349.

Etlinger, Edmund, 18(30)-97, telegraph company director in Spain.
    Elec - 39, 35.
    Elec Rev - 40, 645*.

Evans, Frederick John, 1818-80, gas light manufacturer.
    PICE - 63, 311-313.

Evans, J   , 18(33)-82, dock engineer, Cardiff.
    Engng - 33, 410.

Evans, John, F.R.S., 1818-72, railway engineer.
    PICE - 38, 309-310.

Evans, Richard,   -1894, company secretary.
    JISI - 45, 391*.

Evans, T   F   ,   -1891, Inspector of Mines.
    Iron - 37, 493.

Evans, Thomas,   -1847, iron works manager.
    PICE - 7, 15*.

Evans, Thomas, 1829-87, Inspector of Mines.
    TCMCIE - 16, 11-12.

Evans, William,   -1899, steel manufacturer.
    JISI - 55, 261.

Evelegh, Lewis Frederick, 1858-93, railway engr overseas.
    PICE - 115, 399-400.

Evington, Herbert, 1865-98, draughtsman.
    PICE - 135, 366.

Ewart, Peter, 1767-1843, Chief Inspector of Machinery for Navy.
    CEAJ - 6, 102-103.

Ewing, Ludovic Stewart Rudolph, 1856-86, rly engr overseas.
    PICE - 88, 444-445.

Exall, William, 1808-81, ironmaster & inventor(B).
    Eng - 52, 67.
    PICE - 67, 405-407.

Faber, Charles Wilson, 1813-78, railway director.
   PICE - 55, 330*.

Faija, Henry, 1844-94, cement manufacturer.
   Eng - 78, 189.
   Engng - 58, 297.
   PICE - 119, 391-392.
   TSE - 1894, 256-257.

Fairbairn, Sir Peter, 1797-1861, engineering manufr, inventor(B,D).
   Eng - 11, 29.

Fairbairn, Sir William, F.R.S., 1789-1874, mechanical & manufact-
uring engineer(B,D).
   Eng - 38, 154; 44, 19,38,57,95,111,163,181,186,240,253,291,309.
   Engng - 18, 147-148.
   Iron - 4, 237.
   JISI - 5, 484.
   PICE - 39, 251-264.
   PIME - 1875, 22-24.
   TINA - 16, 263-265.

Fairholme, Capt. Charles, R.N., 1829-91, inventor.
   Eng - 72, 323.
   Engng - 52, 456.
   Iron - 38, 341.
   PIME - 1891, 473.

Fairlie, Robert Francis, 1831-85, railway engineer(B,D).
   Eng - 60, 109.
   Engng - 40, 133.
   Iron - 26, 137.

Falkiner, Travers Hartley, 1829-97, rly contracting engr.
   PICE - 131, 364-367.

Falshaw, Sir James, Bart., 1810-89, civil engng contractor(B).
   Engng - 47, 696.
   Iron - 33, 539.
   PICE - 99, 382-390.

Farey, Barnard William, 1827-88, inventor.
   Engng - 45, 574.
   PICE - 94, 298-300.

Farey, John, 1791-1851, consulting civil engineer(B,D).
   PICE - 11, 100-102.

Farley, Reuben, 1826-99, colliery owner.
   Eng - 87, 271*.
   JISI - 55, 261-262.

Farmer, Christopher Atkinson, 18(44)-95, engineering manufr.
   TMAE - 5, 265.

Farmer, Sir James, 1823-92, engineering manufacturer(B).
    Iron - 40, 384*.
    TMAE - 2, 278-279.

Farmer, John, 1796-1863, ironworks manager.
    PIME - 1864, 14.

Farmer, John Stimson, 1827-92, railway signal manufacturer(B).
    Engng - 54, 780.
    PICE - 112, 374.
    TSE - 1892, 257-258.

Farnworth, William, 18(28)-97, steel works manager.
    JISI - 51, 310-311.

Farquharson, John, 18(22)-98, admiralty electrician.
    Elec Rev - 42, 51.

Farrar, James, 1825-89, surveyor & civil engineer.
    PICE - 98, 390-391.
    PIMCE - 16, 276*.

Farrell, James Barry, 1810-93, municipal surveyor.
    PICE - 113, 344-345.

Faulkner, John, 18(23)-97, electrician.
    Elec Rev - 40, 218.
    Eng - 83, 163.

Faviell, Samuel Clough, 1835-74, company representative.
    PIME - 1875, 24.

Fawcett, Thomas Constantine, 1839-99, machinery manufacturer.
    PIME - 1899, 615.

Fawcus, Capt.William,   -1892, cable ship captain.
    Elec - 29, 54*.

Fearfield, John Piggin, 1850-92, engineering manufacturer.
    PIME - 1892, 98.

Fellows, Samuel James,   -1895, iron plate manufacturer.
    JISI - 47, 261*.

Fenton, James, 1815-63, consulting mechanical engineer.
    Eng - 15, 266*.
    PICE - 23, 487-488.
    PIME - 1864, 14-15.

Fenton, James, 1848-93, rly inspector in India.
    PIME - 1893, 493.

Fenwick, John, 1826-96, mechanical engineer.
    PICE - 126, 401-402.

Ferguson,   John, 1823-87, shipbuilder.
   Engng - 43, 567.
   TIESS - 30, 310-311.

Ferguson, John Dunn, 1864-94, mechanical engineer.
   PICE - 120, 371.

Fewtrell, Samuel,   -1893, ironworks manager.
   Iron - 41, 143.

Field, Arthur, 1844-71, sanitary engineer.
   PICE - 33, 259-261.

Field, Joshua, F.R.S., 1786-1863, marine engineer(B,D).
   Eng - 16, 112.
   PICE - 23, 488-492.
   PIME - 1864, 15-16.

Field, Rogers, 1831-1900, hydraulic & drainage engineer(B).
   Eng - 89, 357*, 362.
   Engng - 69, 445.
   PICE - 141, 343-344.

Fincham, John, 1784-1859, marine engineer.
   Eng - 8, 414*.

Fincken, Christopher Williams,1849-95, colliery manager.
   JISI - 47,261*.

Findlay, Edward Walton, 1848-98, explosives expert.
   Engng - 66, 832.
   TIESS - 42, 403-405.

Findlay, Sir George, 1829-93, railway manager(B).
   Eng - 75, 277.
   Engng - 55, 390.
   Iron - 41, 276.
   PICE - 113, 362-371.

Finlayson, John,   -1893, works manager.
   Iron - 41, 339.

Firbank, Ralph, 1837-82, railway contractor.
   PICE - 72, 317-319.

Firth, Charles Henry,   -1892, steel manufacturer.
   Iron - 40, 514.

Firth, J   Loxley, 18(58)-97, manufacturer.
   Eng - 84, 652.

Firth, Mark, 1819-80, steel manufacturer(B,D).
   Eng - 50, 417.

JISI - 17, 687-689.
Prac Mag - 6, 289-292.

Firth, William, 1835-92, iron & steel manufacturer.
PIME - 1892, 404.

Fishe, Charles Caulfield, 1821-70, railway engineer.
PICE - 31, 218-219.

Fishenden, Frederic George, 1832-96, admiralty engineer.
PICE - 126, 396-397.

Fisher, Benjamin James, 1838-97, railway engineer.
PICE - 128, 344.

Fisher, Benjamin Samuel, 1836-83, locomotive superintendent.
PIME - 1884, 62-63.

Fisher, George, 1809-91, superintendent engineer.
Eng - 71, 380.
Engng - 51, 553*.
PICE - 107, 398-400.

Fisher, Joseph, 1832-77, engineering contractor.
PICE - 50, 186.

Fitze, W   J   ,   -1889, naval engineer.
Eng - 68, 428.
Iron - 34, 445.

Fitzgerald, George Francis, 1851-1900, Prof. of Natural Philosophy, Trinity College, Dublin.
JIEE - 30, 1244-1246.

Fitz-Gibbon, Abraham Coates, 1823-87, civil engineer(B).
PICE - 89, 466-470.

Flanagan, Terence Woulfe, 1819-59, railway engineer.
PICE - 20, 137-138.

Fletcher, David Hardman, 1847-84, tool manufacturer.
PIME - 1885, 72.

Fletcher, Edward, 1807-89, locomotive superintendent.
Eng - 69,18.
Engng - 49, 11.
PIME - 1889, 748-749.

Fletcher, Edward, 1840-87, tool manufacturer.
PIME - 1887, 467.

Fletcher, George, 1837-97, mechanical engineer.
PIME - 1897, 134.

78

Fletcher, Henry Alcock, 1820-61, railwayman in India.
  PICE - 21, 585-586.

Fletcher, Henry Allason, 1834-84, steam engine manufacturer.
  PICE - 78, 417-418.
  PIME - 1884, 472-473.

Fletcher, Herbert, 1842-95, colliery engineer.
  PIME - 1895, 533.
  TFIME - 12, 125-126.

Fletcher, James, 1806-81, tool manufacturer.
  Eng - 51, 217.
  PIME - 1882, 5-6.

Fletcher, James, Jun., 1838-74, tool manufacturer.
  PIME - 1875, 24-25.

Fletcher, Lavington Evans, 1822-97, consulting mechanical engr.
  Eng - 83, 614*, 646; 84, 65+port.
  Engng - 63, 827, 859-860.
  PICE - 129, 368-371.
  PIME - 1897, 134-136.

Fletcher, William, 1831-1900, coal & ironmaster(B).
  JISI - 58, 388-389.

Fletcher, William F   , 18(27)-91, company representative.
  Iron - 38, 341.

Flower, James John Alexander, 1847-88, company representative.
  PICE - 95, 384-385.

Flower, Philip William, 1848-92, tinplate manufacturer.
  Iron - 39, 298.
  JISI - 42, 291.

Floyer, George Wadham, 1863-90, railway engineer overseas.
  PICE - 103, 383-384.

Flynn, Patrick John, 1838-94, Public Works engr in India.
  PICE - 118, 445-446.

Fogg, Robert, 1822-97, consulting mechanical engineer.
  PICE - 130, 313-314.
  PIME - 1897, 233.

Foggin, William, 1850-98, mining engineer.
  PICE - 138, 490.

Folkard, Arthur Crouch, 1836-96, Public Works engr in Ceylon.
  Eng - 83, 5.
  PICE - 127, 364-365.

Foord, Alfred Montague, 1858-94, Public Works engr in India.
PICE - 116, 381.

Forbes, David, F.R.S., 1828-76, metallurgist & geologist(B,D).
JISI - 9, 519-524.
PICE - 49, 270-275.

Forbes, Maj.-Gen. William Nairn, 1796-1855, Bengal Engineer(B).
PICE - 20, 138-140.

Forde, Arthur William, 1821-86, rly contractor in India.
PICE - 88, 436-437.

Forde, Henry Charles, 1827-97, telegraph engineer(B).
Elec - 38, 579.
Elec Rev - 40, 281.
Eng - 83, 213.
Engng - 63, 287-288.
PICE - 128, 345-346.

Forman, James Richardson, 18(22)-1900, civil engineer.
Eng - 90, 39.
Engng - 70, 52.
PICE - 142, 356.

Forrest, Alfred, 1864-94, railway engineer overseas.
PICE - 118, 459-460.

Forrest, Jacob, 1830-97, coal & ironmaster.
PICE - 130, 314.

Forrest, William John, 1828-73, civil engineer.
Eng - 16, 315*.
PICE - 38, 286-287.
PIME - 1874, 19.

Forster, Edward, 1813-87, glass works manager.
PIME - 1888, 154-155.

Forster, Frank, 1800-52, sanitary engineer(B).
PICE - 12, 157-159.

Forster, Thomas Emerson, 1802-75, colliery viewer(B).
PICE - 43, 300-303.
TNEIMME - 25, 5-10.By G.C. Greenwell.

Forsyth, John Curphey, 1815-79, railway engineer.
PICE - 58, 343-345.

Fortescue, Hugh Ebrington, 1825-59, railway engineer.
PICE - 19, 173-174.

Foster, Frederick, 1840-93, manufacturer.
PIME - 1893, 493-494.

Foster, John Francis, 1867-99, civil engineer.
PICE - 137, 440-441.

Foster, Samuel Lloyd, 1831-79, iron & coal company director.
PIME - 1880, 3.

Foster, William Orme, 1814-99, ironmaster(B).
Eng - 88, 358.
JISI - 56, 293.

Fothergill, Thomas,   -1858, ironmaster.
Eng - 5, 448.

Fouracres, Charles, 18(26)-84, Public Works engr in India(B).
PICE - 78, 418-424.

Fowke, Capt.Francis, 1823-65, Royal Engineer(B,D).
PICE - 30, 468-470.

Fowler, Francis, 1829-93, railway engineer overseas.
PICE - 114, 372-374.

Fowler, Henry, 1821-54, railway engineer.
PICE - 14, 133-135.

Fowler, Sir John, Bart., 1817-98, civil engineer(B,D).
Elec Rev - 43, 795.
Eng - 86, 513+port.
Engng - 66, 688-690+port.
PICE - 135, 328-337.
PIME - 1899, 128-129.
Prac Mag - 5, 257-262.
TSE - 1898, 239-240.

Fowler, John, 1824-88, river engineer.
Engng - 46, 389.
Iron - 32, 353.
PICE - 95, 371-374.
Prac Mag - 1875, 257-262+port.
TSE - 1868, 299-318.

Fowler, John, 1826-64, agricultural engineer(B,D).
PIME - 1865, 14-15.
Prac Mag - 1875, 257-262 +port.
TSE - 1868, 299-318.

Fowler, Robert, 18(24)-88, steam plough manufacturer.
Iron - 32, 503.

Fowler, William, 1820-76, iron & coal master.
JISI - 9, 511-512.

Fowls, Samuel, 1779-1848, river engineer.
PICE - 8, 11.

Fox, Sir Charles, 1810-74, civil & consulting engr(B,D).
Eng - 37, 404.
Engng - 18, 53.
PICE - 39, 264-266.
Prac Mag - 6, 129-133+port.

Fox, Samuel, 1815-87, umbrella frame manufacturer(B).
Eng - 63, 171.
Iron - 29, 185.

Fox, Theodore, 18(31)-99, manufacturer.
Eng - 88, 359.

Fox, William, 1840-91, tool manufacturer.
Iron - 37, 228.
PIME - 1891, 191-192.

France, William, 1840-87, mining engineer.
JISI - 32, 213-214.

Francis, Charles Larkin, 1801-73, cement manufacturer(B).
PICE - 38, 311.

Francis, John,   -1894, civil engineer.
Engng - 58, 609.

Francis, William, 1831-88, surveyor & railway engineer.
PICE - 95, 374-375.

Franks, Thomas, 18(28)-92, iron merchant.
Iron - 40, 362*.

Fraser, Alexander, 1823-95, waterworks engineer.
Eng - 80, 414+port.
PICE - 123, 433-434.

Fraser, Alfred, 1853-96, waterworks engineer.
PICE - 125, 403-404.

Fraser, Alister, 1829-69, railway engineer.
PICE - 31, 239-240.

Fraser, Capt.Edward, 1822-57, Bengal Engineer.
PICE - 20, 163.

Fraser, Henry John, 1848-89, railway contractor.
PICE - 100, 383-384.

Fraser, John, 1819-81, railway engineer.
Eng - 52, 247.
PICE - 70, 417-419.
PIME - 1882, 6-7.

Fraser, Robert Samuel, 1829-84, armaments engineer.
PICE - 78, 424-429.

Frazer, John, 1833-95, superintendent engineer.
    TIESS - 39, 269-270.

Freeman, John Russell, 1826-78, stone merchant.
    PICE - 52, 281-282*.

Freeman, Joseph, 1819-71, company representative.
    Engng - 11, 160.
    PICE - 33, 261*.
    PIME - 1872, 16-17.

Freeman, Samuel Tate, 1827-71, railway engineer.
    PICE - 33, 261-263.

Freeman, William, 1793-1880, stone merchant.
    PICE - 63, 326-327.

French, Henry, 18(30)-97, colliery manager.
    Eng - 83, 525.

French, James, 1846-94, contracting & mining engineer.
    PICE - 119, 406-407.

Frend, William Arthur, 1858-89, rly engineer in South America.
    PICE - 98, 403-404.

Frew, John, 18(49)-99, ironworks manager.
    JISI - 56, 293-294.

Frewer, Charles, 1866-85, metal dealer.
    PICE - 82, 389-390.

Frith, E   , 18(18)-89, engr to Duke of Devonshire.
    Eng - 67, 401.

Frodsham, Charles, 1810-71, chronometer manufacturer.
    PICE - 33, 263-264.

Frome, Gen. Edward, 1802-90, Royal Engineer(B).
    PICE - 101, 284-285.

Froude, William, F.R.S., 1810-79, naval architect(B,D).
    Eng - 47, 395.
    Engng - 27, 462, 503, 525; 28, 76.
    Iron - 13, 692.
    PICE - 60, 395-404.
    PIME - 1881, 3-5.
    TINA - 20, 264-269.

Fry, Edward, 1839-89, municipal engineer.
    PIMCE - 16, 275-276.

Fryer, Alfred, 1830-92, inventor, manufacturer(B).
    Engng - 54, 780.

Fulcher, George Chambers, 1868-1900, draughtsman.
    PIME - 1900-1, 469.

Fuller, J   W   , 18(43)-89, telegraph engineer.
    Elec - 22, 653.

Fulton, Hamilton Henry, 1813-86, civil engineer(B).
    Eng - 62, 158.
    Engng - 42, 185.
    PICE - 87, 418-422.

Fulton, John,   -1891, municipal engineer.
    Iron - 37, 537*.

Furness, Edward, 1843-99, consulting mechanical engineer.
    PIME - 1899, 615-616.

Furness, George, 1820-1900, contractor(B).
    Eng - 89, 78.
    PICE - 140, 286-287.

Furness, Henry David, 1830-80, consulting engineer.
    PICE - 63, 313-314.

Fyfe, Alexander L   ,   -1899, electrical engineer.
    Elec Rev - 45, 606*.

Galbraith, Ewan Edward, 18(77)-98, ---
    Eng - 86, 375*.

Gale, Henry, 1836-98, railway engineer.
    PICE - 133, 398-399.

Galland, Nicolas,   -1886, brewing engineer.
    Eng - 61, 365.

Gallow, Thomas, 18(37)-96, mechanical engr, inventor.
    Eng - 81, 63.

Galloway, John, 1804-94, boiler manufacturer(B).
    Elec - 32, 406.
    Eng - 77, 129+port, 190.
    Engng - 57, 227.

Galloway, John, Jun., 1825-96, manufacturing engineer.
    Elec Rev - 39, 805*.
    Eng - 82, 646; 83, 9+port.
    Engng - 62, 771.
    JISI - 51, 311.
    TMAE - 6, 318-320.

Galloway, John Alexander, 1804-50, consulting engineer.
    PICE - 10, 81-82.

Gallwey, Lionel Philip Payne, 1851-91, engr in West Indies.
    Eng - 72, 323.
    PICE - 107, 400-402.

Galsworthy, J   ,   -1880, municipal engineer.
    PIMCE - 7, 147*.

Galton, Capt. Sir Douglas Strutt, F.R.S., 1822-99, Royal
    Engineer, railway & sanitary engineer(B,D).
    Elec - 42, 725.
    Elec Rev - 44, 427.
    Eng - 87, 257+ port.
    Engng - 67, 353-354+port.
    JIEE - 28, 674-675.
    JISI - 55 , 262.
    PICE - 137, 413-417.
    PIME - 1899, 129-134.

Galwey, William, 1828-76, railway engineer.
    Iron - 38, 341.
    PICE - 49, 259-261.

Galwey, William St.John, 1833-91, rly engr in Far East.
    Eng - 72, 323.
    PICE - 109, 399-403.

Gamble, John George, 1842-89, hydraulic engineer(B).
    Eng - 68, 416.
    Iron - 34, 425.
    PICE - 111, 373-378.

Gammon, William, 1841-68, railway contractor.
    PICE - 31, 240-241.

Garbutt, Henry, 1871-1900, mechanical engineer.
    PIME - 1900, 625.

Gardiner, Robert Barlow, 1818-59, surveyor.
    PICE - 19, 188.

Gardner, John, 1821-94, railway surveyor.
    PICE - 116, 358.

Gardner, Thomas William, 1841-(71), hydraulic engineer.
PICE - 33, 264-265.

Garland, Joseph, 1805-91, shipbuilder.
Engng - 51, 565.

Garland, Joseph, 1840-99, mining engineer.
PICE - 139, 354-356.

Garland, Thomas Bland, 1819-92, engr in South America.
PICE - 109, 426-428.

Garlick, Edward, 1822-1900, municipal engineer.
PICE - 142, 356-357.

Garnett, George, 1820-94, gas works manager.
PICE - 120, 354-355.
TSE - 1894, 258.

Garrett, Gilbert Henry, 1855-89, engng works manager.
NECIES - 5, xliii.
PICE - 97, 421.

Garrett, Richard, 18(06)-66, agricultural engineer.
Eng - 22, 8.

Garrett, Richard, Jun., 1829-84, agricultural machinery manufr(B
Eng - 58, 109-110.
Engng - 38, 178.
Iron - 24, 148.
PICE - 78, 429-431.
PIME - 1884, 400-401.

Garrett, Robert, 1823-57, railway engineer in India.
PICE - 18, 188-189.

Gateshead, James, 18(44)-96, civil engineer.
Engng - 62, 561-562.

Gauntlett, William Henry, 1823-98, iron & steel manufacturer.
Eng - 86, 406*.
PIME - 1898, 703-704.

Gavin, T   , 1826-89, railway contractor.
Engng - 48, 136*.

Gaynor, Capt.Henry Francis, 1864-98, Royal Engineer.
PIME - 1899, 266-267.

Geach, Charles, M.P., 1808-54, engineering manufacturer.
PICE - 14, 148-151.

Geach, John J   , 1840-83, civil engineer.
PIME - 1884, 401*.

Geddie, James, 18(19)-95, shipbuilder.
   Engng - 60, 124.

Gedhill, Tom, 18(30)-93, municipal surveyor.
   PIMCE - 19, 365.

Gee, Alfred,   -1858, civil engineer.
   Eng - 6, 191*.

Geneste, Frank Alexander Brown, 1842-88, rly engr overseas.
   PICE - 95, 375-377.

Gent, John Thomas, 18(42)-98, electrical manufr, inventor.
   Elec Rev - 43, 283.

George, John Rees,   -1889, engineer in New Zealand.
   PICE - 99, 374-376.

George, Robert John, 1841-90, railway engineer.
   PICE - 103, 366.

Gibb, Alexander, 1804-67, locomotive superintendent(B).
   Mech Mag - 87, 115.
   PICE - 27, 587-589.

Gibb, John, 1776-1850, contractor, civil engineer(D).
   PICE - 10, 82-85.

Gibbons, Benjamin, 1783-1873, ironmaster.
   PIME - 1874, 19-20.

Gibbons, Benjamin, Jun., 1815-63, ironmaster.
   PIME - 1864, 16.

Gibbons, William Coulthurst, 1852-81, Public Works engr in India.
   PICE - 66, 382-383.

Gibbs, John Douglas Lyle,   -1891, telegraph engineer.
   Elec - 27, 599*.

Gibbs, Joseph, 1789-1864, mechanical engineer, inventor(B).
   PICE - 24, 528-531.

Gibson, George Henry, 1852-99, contractor.
   PIME - 1900, 331-332.

Gibson, Thomas, 1843-99, contracting engineer.
   PICE - 140, 282.

Gilbert, Charles,   -1900, publisher of "Engineering".
   Engng - 69, 98.

Gilbert, Davies, F.R.S.,   -1840, mechanical engineer.
   CEAJ - 3, 66-67.

Gilbert, Edward,   -1891, telegraph engineer.
    Elec - 27, 459.

Gilchrist, Archibald, 1822-1900, shipbuilder.
    Eng - 89, 44.
    Engng - 69, 53.
    TIESS - 43, 359.

Gilchrist, William Gilchrist, 1849-99, railway engr in India.
    PICE - 137, 423-424.

Giles, Alfred, 1816-95, civil engineer(B).
    Eng - 79, 199+port.
    Engng - 59, 316-317.
    PICE - 122, 367-372.

Giles, Edward, 1842-78, railway engineer overseas.
    PICE - 53, 289.

Giles, Francis, 1806-84, railway & dock engineer.
    PICE - 80, 332.

Giles, Francis John William Thomas, 1787-1847, civil engineer(D).
    PICE - 7, 9.

Giles, George, 1810-77, railway engineer.
    PICE - 50, 177-178.

Gilkes, Edgar, 1821-94, manufacturing engineer.
    PIME - 1895, 142-143.

Gill, Henry, 1824-93, Berlin waterworks engineer.
    PICE - 114, 374-376.

Gill, Robert, 1846-92, mechanical engineer.
    Eng - 73, 348.

Gillespie, W   H   ,   -1875, coal owner.
    Engng - 19, 225.

Gillham, Francis, 1853-92, rly engr in Central America.
    PICE - 110, 388-389.

Gilroy, George, 1823-94, consulting mining engineer.
    PICE - 116, 359.

Gimingham, Charles Henry, 1853-90, electric light company director
    Elec - 25, 625.

Gimson, Josiah, 1818-83, machinery manufacturer.
    PIME - 1884, 63-64.

Ginty, William Gilbert, 1820-66, gas engr & surveyor.
    PICE - 27, 589-590.

Girdlestone, Henry John, 1824-97, manufacturer.
    PICE - 129, 393.

Gisborne, Lionel, 1823-61, civil engineer(B).
    PICE - 21, 586-592.

Gjers, John, 1830-98, ironmaster.
    Eng - 86, 398-399.
    Engng - 66, 496.
    JISI - 54, 327-328.
    PIME - 1898, 704-706.
    TSE - 1898, 238-239.

Glass, Sir Richard Atwood, 1820-73, cable manufacturer(B,D).
    Eng - 37, 4*.
    PICE - 42, 262.

Gledhill, John,   -1892, municipal surveyor.
    PIMCE - 19, 365*.

Gledhill, Manassah, 1826-98, engineering manufacturer.
    Eng - 86, 279.
    PIME - 1899, 267-269.
    TMAE - 8, 220-221.

Glen, David Corse, 1824-92, steam hammer manufacturer.
    Engng - 54, 643.
    TIESS - 36, 318.

Glinn, George James Harvey, 1829-75, civil engineer.
    PICE - 41, 216-217.

Glover, Walter T   , 1846-93, electric wire manufacturer.
    Elec - 31, 4.
    Elec Rev - 32, 529.

Glynn, Joseph, F.R.S., 1799-1863, mechanical & civil engr(B).
    Eng - 15, 98-99.
    PICE - 23, 492-498.

Godber, Samuel, 1845-86, colliery engineer.
    TCMCIE - 15, 32.

Goddard, Ebenezer, 1816-82, gas engineer.
    Iron - 20, 360.
    PICE - 72, 312-314.

Godfrey, Samuel, 1832-87, mechanical engineer.
    Iron - 29, 343.
    PIME - 1887, 275-277.

Godson, William F   , 1807-67, railway superintendent.
    Eng - 25, 2*.

Godwin, Henry Colthurst, 1858-92, railway engr overseas.
PICE - 112, 367.

Godwin, John,   -1869, railway engineer(B).
PICE - 30, 434-435.

Going, Thomas Hardinge, 1827-75, engineer in India.
Engng - 20, 496.
PICE - 44, 221.

Gooch, Alfred William, 1846-87, superintendent rly engineer.
Engng - 43, 530*.

Gooch, Sir Daniel, Bart., 1816-89, railway chairman(B,D).
Elec - 23, 601, 618.
Eng - 68, 335.
Engng - 48, 466.
Iron - 34, 337.

Gooch, John Viret,   1812-1900, locomotive superintendent.
Engng - 69, 789.
PICE - 141, 344-347.

Gooch, Thomas Longridge, 1808-82, railway engineer(B).
Iron - 20, 464.
PICE - 72, 300-308.

Goodall, Hamilton, 1848-87, boiler manufacturer.
PICE - 90, 439-441.

Goodfellow, Benjamin, 1811-63, engineering manufacturer.
Eng - 15, 266.
PIME - 1864, 16-17.

Goodfellow, Charles Edward, 1855-99, civil engineer.
PICE - 138, 497.

Goodfellow, Joseph, 1841-99, railway engineer.
Eng - 88, 17-18.
TIESS - 42, 405-406.

Gooding, William Oliver, 1837-75, civil engineer.
PICE - 45, 251-252.

Goodman, John Brooke, 1874-98, electrician.
Elec Rev - 42, 812.

Goodwin, Gilbert, 18(71)-95, marine engineer.
TIESS - 39, 274.

Goolden, Joseph Henry, 1853-92, municipal surveyor.
TSE - 1892, 257.

Gordon, Alexander, 1802-68, civil engineer.
PICE - 30, 435-436.

Gordon, James Edward Henry, 1852-93, consulting electrical engr(B).
Elec - 30, 417-418.
Elec Rev - 32, 159-160.
Eng - 75, 131.
Engng - 55, 119.
PICE - 113, 346-347.

Gordon, Maj.-Gen. Sir John, 1805-70, Royal Engineer.
PICE - 31, 241-245.

Gordon, Joseph, 1836-89, municipal engineer.
Eng - 68, 418.
Iron -34, 425.
PICE - 100, 384-387.
PIMCE - 16, 270-275+port frontis.

Gorman, William, 1811-85, iron & steel manufacturer.
Engng - 40, 548.
Iron - 26, 485.

Gostling, Herbert Charles, 1868-98, rly engr in S.America.
PICE - 133, 414.

Gotto, Edward, 1822-97, consulting drainage engineer.
PICE - 128, 346-347.

Gough, Nathan, 1790-1852, engineering manufacturer.
PICE - 14, 152.

Gowenlock, Alfred Hargreaves, 18(32)-82, iron founder in India.
PIME - 1883, 18.

Graham, Edwin, 1837-97, shipbuilder.
NECIES - 13, 270.

Graham, George, 1822-99, railway engineer.
Eng - 88, 7-8+port.
Engng - 68, 24+port.
PICE - 137, 424-427.
TIESS - 42, 406-408.

Graham, James, 1827-78, surveyor.
PICE - 56, 284-286.

Grahame, William Henry, 1858-95, engineer in New Zealand.
PICE - 122, 397.

Grainger, James Nixon, 1847-81, engineer in India.
PIME - 1882, 8.

Grainger, Thomas, 1794-1852, civil engineer(B).
    PICE - 12, 159-161.

Grant, Alfred,   -1897, surveyor.
    Eng - 83, 95.

Grant, John, 1819-88, municipal engineer(B).
    Eng - 65, 283.
    Engng - 45, 359.
    PICE - 92, 389-392.

Grant, John,   -1897, shipbuilder.
    Engng - 64, 495.

Grant, John Duncan, 1848-93, Public Works engr in India.
    PICE - 116, 360-361.

Grant, John Maitland, 1851-94, rly surveyor & engineer.
    PICE - 118, 446-447.

Grant, Thomas Barrett, 18(62)-95, electrical manufacturer.
    Elec - 34, 388*.
    Elec Rev - 36, 126*.

Grantham, John, 1809-74, consulting marine engineer(B).
    Eng - 38, 80.
    Engng - 18, 75.
    PICE - 39, 266-268.
    TINA - 16, 270-272.

Grantham, Richard Boxall, 1805-91, consulting civil engr(B).
    Eng - 72, 483.
    Iron - 38, 558.
    PICE - 108, 399-403.

Granville, 2nd Earl of, 1815-91, ironmaster & coal owner(B).
    JISI - 39, 232-234.
    PICE - 104, 287-288.
    TSE - 1891, 215-216.

Gravatt, William, F.R.S., 1806-66, civil engineer(B).
    Elec Rev - 31, 575, 610*.
    Eng - 21, 421*.
    PICE - 26, 565-575.

Graves,   ,   -1897, telegraph engineer in Persia.
    Elec Rev - 41, 837.

Graves, Edward, 1834-92, telegraph engineer.
    Elec - 30, 39.
    Iron - 40, 427.

Gray, Archibald,   -1884, ---
    TIESS - 28, 290*.

Gray, John Kaye, 18(56)-89, submarine telegraph engineer.
    Elec Rev - 25, 582.

Gray, John William, 1828-96, water works engineer.
    Eng - 82, 189.
    PICE - 126, 397-398.
    PIME - 1896, 256-258.

Gray, Matthew, 1856-96, steel manufacturer.
    JISI - 50, 258.

Gray, Richard Armstrong,   -1899, municipal engineer.
    PIMCE - 25, 479*.

Gray, Thomas, 18(32)-90, Secretary, Marine Department, Board of
    Trade.
    Eng - 69, 243, 255.
    Engng - 49, 370.
    Iron - 35, 225-226.
    TIESS - 33, 209-210.
    TINA - 31, 289-290.

Gray, Sir William, 1823-98, shipbuilder(B).
    Eng - 86, 278.
    Engng - 66, 368.
    NECIES - 15, 263-265.

Grazebrook, Henry, 1818-83, engineering works manager.
    JISI - 23, 670-671.

Greathead, James Henry, 1844-96, railway engineer(B).
    Elec - 37, 817*.
    Elec Rev - 39, 573.
    Eng - 82, 428, 448 +port.
    JISI - 49, 258-259.
    PICE - 127, 365-369.
    PIME - 1896, 598-599.

Greaves, Charles, 1816-83, waterworks engineer(B).
    Eng - 56, 358.
    Iron - 22, 422.
    PICE - 76, 355-359.

Greaves, James Henry, 1846-79, steel works manager in Germany.
    PICE - 56, 289.
    PIME - 1880, 285.

Greck, Col. Peter, 1827-88, engr in Russia.
    JISI - 34, 222.
    PICE - 94, 300-301.

Green, Charles, 1798-1866, boiler tube manufacturer.
    PIME - 1867, 14-15.

Green, Charles Frederic, 1845-86, surveyor.
    PICE - 88, 437-438.

Green, George, 1796-1849, shipbuilder.
    Eng - 87, 81+port.

Green, James, 1781-1849, sanitary engineer.
    PICE - 9, 98-100.

Green, John, 1787-1852, civil engr & architect(B).
    PICE - 13, 138-140.

Green, Robert Wood Everett, 1847-76, engineer in India.
    PICE - 48, 270-272.

Green, Samuel, -1887, mining engineer.
    Eng - 63, 415.

Green, T L , -1898, file manufacturer.
    Engng - 65, 144.

Greene, Robert Rowan, 1835-82, railway engineer.
    PICE - 72, 308-309.

Greener, John Henry, 1829-95, telegraph engineer.
    Elec - 34, 733-734.
    Elec Rev - 36, 454*.
    Eng - 79, 311.
    PICE - 122, 407-408.
    PIME - 1895, 310-311.

Greenwell, Edwin Walter, 1867-96, harbour engineer.
    PICE - 127, 389.

Greenwell, George Clementson, 1821-1900, consulting mining engr.
    Eng - 90, 467.
    PICE - 143, 316-317.
    TIME - 22, 124-126+port. By G.C. Greenwell.

Greenwood, George, 1840-98, machinery manufacturer.
    Elec Rev - 42, 587.
    Engng - 65, 497.
    PICE - 133, 399-400.

Greenwood, Thomas, 1809-73, engineering manufacturer(B).
    Eng - 35, 113.
    Engng - 15, 148-149.
    PICE - 38, 311.
    PIME - 1874, 20-21.

Greenwood, Thomas, 1833-73, small arms manufacturer.
    PICE - 38, 311-313.

Gregory, Sir Charles Hutton, 1817-98, consulting civil engineer(B
    Elec Rev - 42, 51.

94

Eng - 85, 39.
Engng - 65, 52-53.
PICE - 132, 377-382.

Greig, David, 1827-91, steam plough manufacturer.
Eng - 71, 248-249.
Engng - 51, 378, 407.
Iron - 37, 297.
PIME - 1891, 474-476.

Greig, Maj.-Gen. Irwin Montgomery, 1834-87, Bombay Engineer(B).
PICE - 90, 449-450.

Grew, Nathaniel, 1829-97, consulting mechanical engineer.
PICE - 130, 315.
PIME - 1897, 233-234.

Grice, Edwin James, 1834-89, nut & bolt manufacturer.
JISI - 34, 221-222.
PICE - 96, 318-319.
PIME - 1889, 333-334.

Grice, Frederick Groom, 1829-81, ironmaster.
PIME - 1882, 8.

Grier, William Magee, 1839-93, Public Works engr in S.Africa.
PICE - 114, 376-378.

Grierson, James, 1827-87, railway company manager(B).
Engng - 44, 417.
Iron - 30, 353.
PICE - 99, 390-397.

Grieve, Charles Archibald, 1851-79, railway engineer.
PICE - 60, 406-407.

Griffin, Richard T   , 18(47)-94, technical publisher.
Eng - 77, 287*.

Griffith, Sir Richard, Bart., 1784-1878, Inspector of Mines in
Ireland(B).
Eng - 46, 220.
Iron - 12, 396.
PICE - 55, 317-318.

Griffith, William Charles Easton, 1854-93, civil engr overseas.
PICE - 113, 347-349.

Griffiths, James Evans, 1846-99, consulting mechanical engr.
PIME - 1899, 471.

Griffiths, John,   -1880, contractor.
Engng - 29, 342*.

95

Griffiths, Robert, 1805-83, inventor of screw propeller(B,D).
    Eng - 55, 477-478.
    Engng - 35, 606.
    Iron - 21, 562.

Griffiths, Samuel,   -1881, engineering journalist.
    Engng - 31, 549.

Grissell, Henry, 1817-83, iron founder.
    PICE - 73, 376-378.

Grissell, Thomas, 1801-74, contractor(B).
    Eng - 37, 364*.
    Engng - 17, 428.
    PICE - 39, 289-290.

Grose, Daniel Gallagher, 1828-68, railway engineer.
    PICE - 30, 436-437.

Gross, S   , 1791-1866, mining engineer.
    Eng - 21, 470*.

Grove, Sir William Robert, F.R.S., 1811-96, electrical engr(B,D).
    Elec - 37, 483.
    Elec Rev - 39, 181.
    Eng - 82, 143.
    PICE - 127, 358-360.

Grover, Col. George Edward, 1840-93, Royal Engineer.
    PICE - 113, 372-373.

Grover, John William, 1836-92, consulting civil engineer(B,D).
    Eng - 74, 204.
    Engng - 54, 304.
    Iron - 40, 207.
    PICE - 112, 347-349.

Grundy, Ralph Darling, 1846-79, colliery engineer.
    PIME - 1880, 5-6.

Grundy, Robert,   -1859, mechanical engineer in Brazil.
    PICE - 20, 140-141.

Guest, Arthur Edward, 1841-98, railway director(B).
    PICE - 134, 416-419.

Guest, Sir Josiah John, Bart., M.P., 1785-1852, ironmaster(B,D).
    Eng - 23, 595; 26, 243.
    PICE - 12, 163-165.

Guilford, Francis Leaver, 1842-94, works manager.
    PIME - 1894, 160.

Gunter, John,   -1896, ironmaster.
    Eng - 81, 161*.

Guppy, Thomas Richard, 1797-1882, marine superintendent(B).
    Iron - 20, 12.
    PICE - 69, 411-415.

Gurden, Charles Frederick, 1843-74, shipping engineer.
    PIME - 1875, 25-26.

Gurney, Sir Goldsworthy, 1793-1875, inventor(B,D).
    Eng - 39, 184.
    Engng - 19, 212-213.

Gurney, P   ,   -1872, ---
    Engng - 13, 393*.

Guy, Charles William, 1836-93, mechanical engr overseas.
    PIME - 1895, 534.

Hack,Thomas, 1837-88, water works engineer.
    PICE - 93, 488-489.

Hack, William Broughton, 18(06)-75, water works engineer.
    Engng - 20, 473.

Hackford, George, 18(25)-96, railway surveyor.
    Eng - 81, 315.
    Engng - 61, 410.

Hackney, William, 1841-90, steel works manager.
    PICE - 100, 411-412.

Haddan, John Lawton, 1841-80, civil engineer overseas.
    Eng - 49, 256.
    PICE - 62, 352-354.

Haddin, Andrew Aitken, 1849-92, water works engineer.
    PICE - 111, 400.
    TIESS - 36, 319.

Haddon, George Edmunds, 1852-1900, railway engr in India.
    PICE - 143, 318.

Haden, George, 1786-1856, mechanical engineer.
    PICE - 16, 124-127.

Hadfield, Robert, 1831-88, steel manufacturer.
    Iron - 31, 277.
    JISI - 32, 215-216.
    PIME - 1888, 260.
    TCMCIE - 17, 13-18.
    TIESS - 31, 236.

Hadley, Ezra, 18(33)-92, ironmaster.
    Iron - 40, 578.

Haggie, David, 18(19)-95, chain & wire manufacturer.
    Eng - 80, 457.

Haggie, Peter, 1821-86, wire manufacturer.
    PIME - 1887, 146.
    TCMCIE - 16, 12.

Haghe, Augustus, 1846-90, civil engineer.
    PICE - 101, 307-308.

Hague, Joseph, 1843-97, municipal engineer.
    PICE - 131, 383.

Hakewill, Henry, 1842-70, railway engineer.
    PICE - 31, 245-246.

Hall, Charles, 1842-1900, marine engineer.
    PICE - 143, 332.

Hall, George Edward, 1843-98, municipal engineer.
    PIME - 1898, 137-138.

Hall, James, 18(21)-95, manufacturer.
    TMAE - 5, 265.

Hall, James, 18(48)-85, municipal engineer.
    PIMCE - 12, 291.

Hall, Jeffery Austin, 1826-84, municipal engr of Liverpool.
    PIMCE - 10, 310.

Hall, John Francis, 1854-97, steel works manager.
    Eng - 83, 573.
    Engng - 63, 747.
    JISI - 52, 255.
    PICE - 129, 371-372.

Hall, John Green, 18(35)-87, municipal engineer.
    PIMCE - 13, 317.

Hall, Richard, 1806-78, surveyor(B).
PICE - 52, 282.

Hall, Richard Thomas, 1823-89, railway engr in S.Africa.
PICE - 99, 351-352.

Hall, Samuel, 1781-1863, inventor(B,D).
Eng - 16, 314*; 17, 15.

Hall, Sydney, 1813-84, manufacturer(B).
PICE - 79, 366-368.

Hall, Walter,   -1889, electrician & inventor.
Elec Rev - 25, 614*.

Hall, William, 1806-87, shipbuilder.
Eng - 64, 131.
Engng - 44, 248-249.

Hall, William Bancks, 1829-81, railway engineer.
PICE - 65, 366-367.

Hall, William Jeremiah, 1851-90, harbour engineer.
PICE - 102, 334-335.
PIME - 1890, 291.

Halley, David, 18(50)-95, works manager.
Eng - 79, 28.
TIESS - 38, 333-334.

Halliwell, J   ,   -1900, colliery manager.
Eng - 90, 137.

Halpin, Capt.Robert C   , 1836-94, cable ship captain.
Elec - 32, 322.
Elec Rev - 34, 97.

Hamand, Arthur Samuel, 18(39)-88, railway engineer.
PICE - 97, 401-402.

Hambling, Thomas Crump, 1835-73, consulting engineer.
PICE - 38, 313.
PIME - 1874, 21.

Hamill, Edward Dames, 1816-92, contractor.
PICE - 109, 418.

Hamilton, Edward Bell,   -1898, ironmaster.
Eng - 85, 23*.

Hamilton, James, 18(11)-95, shipbuilder.
Engng - 57, 231.

Hamilton, John, 1818-68, manufacturing engineer.
PICE - 30, 470-471.

Hammond, J , -1889, agricultural engineer.
Eng - 68, 371.

Hampson, Robert Stuart, 1860-98, rly telegraph engr.
Elec Rev - 42, 735.
JIEE - 28, 675-676.

Hancock, Thomas, 1845-99, mining engineer overseas.
TSE - 1900, 267.

Hancox, Joseph, 1827-89, railway surveyor.
PICE - 99, 397-398.

Handcock, Robert, -1885, civil engineer overseas.
PICE - 83, 433.

Handyside, James Baird, 18(36)-82, ironworks manager.
Eng - 53, 197.
Engng - 33, 247.
PIME - 1883, 18-19.

Handyside, William, 1793-1850, mechanical engineer in Russia (D)
PICE - 10, 85-87.

Hankey, William Alers, 1771-1859, Treasurer, Institution of
Civil Engineers(B).
PICE - 20, 134-135.

Hanna, Francis Baker, 1841-91, railway engineer in India.
PICE - 109, 403-405.

Hannay, Robert, 1807-74, ironmaster.
PICE - 40, 261.

Hanning, James, 18(30)-93, ---
TMAE - 3, 285*.

Hansen, W , 18(37)-99, manufacturer.
Engng - 67, 616.

Hanson, Conrad Abben, 1837-69, mechanical engineer.
PICE - 31, 246-247.

Hanson, William, 1837-99, iron & coal master.
Eng - 87, 477, 504.
JISI - 55, 262-263.

Hanvey, John, 1824-79, sanitary engineer.
PICE - 59, 314-315.

Hardcastle, William John, 1831-90, lighthouse engr in Middle
    East(B).
    PICE - 97, 402-404.

Harding, John, 1812-71, railway engineer & contractor.
    PIME - 1872, 17.

Harding, Wyndham, F.R.S., 1817-55, railway engineer(B).
    PICE - 15, 97-100.

Hardinge, George, 1831-79, railway engineer in India.
    PICE - 59, 291-293.

Hardy, Hon. Harold Gathorne, 1850-81, ironmaster.
    JIST - 19, 575-576.

Hardy, John, 1820-96, railway engineer.
    Eng - 81, 643; 82, 31.

Hardy, John, 18(24)-1900, iron merchant.
    Eng - 89, 486.

Hardy, Thomas,   -1892, cutlery manufacturer.
    Iron - 40, 514*.

Hardy, William Henry, 1823-95, gas & water works secretary.
    TSE - 1895, 269.

Harford, Edward,   -1898, railway union leader.
    Eng - 85, 11.
    Engng - 65, 58.

Hargreaves, William, 18(21)-89, ironmaster.
    Eng - 68, 295.
    Iron - 34, 318.

Harker, Harold Hayes, 1857-91, locomotive superintendent.
    PICE - 109, 419-420.

Harkness, John McNair,   -1878, surveyor & engineer.
    PICE - 55, 318-319.

Harkness, William, 1810-85, ship builder.
    Engng - 39, 517*.

Harland, Sir Edward James, Bart., 1831-95, ship builder(B).
    Eng - 80, 645; 81, 8+port.
    Engng - 60, 801; 61, 21-24.
    PIME - 1895, 534-537.
    TINA - 37, 397.

Harman, Harry Jones, 1846-95, mechanical engineer.
    PICE - 121, 333-334.

Harman, Henry William, 1815-75, engineering works manager.
PIME - 1876, 20-21.

Harness, Gen. Sir Henry Drury, 1804-83, Royal Engineer(B,D).
PICE - 73, 378-382.

Harper, Edmund Lincoln, 1850-91, engineer overseas.
PICE - 105, 323.

Harpur, Samuel, 1820-88, municipal engineer.
Eng - 66, 406*.
Engng - 46, 480.
PICE - 95, 385-387.

Harris, Edward, 18(16)-1900, civil engineer.
Eng - 90, 215*.

Harris, John, 1812-69, railway contractor.
PICE - 31, 219-220.

Harris, Robert, 1835-92, gas engineer.
PICE - 110, 380.
TSE - 1892, 256-257.

Harris, Sir William Snow, 1791-1867, electrician.
Mech Mag - 86, 55.

Harrison, Frank, 18(60)-98, manufacturer.
Eng - 85, 243*.

Harrison, George, 1815-75, locomotive manufacturer.
PICE - 43, 303-304.
PIME - 1876, 21-22.

Harrison, George, 1830-98, railway engineer.
PICE - 131, 383-384.

Harrison, Henry, 1822-83, contracting engineer.
PICE - 75, 317-318.

Harrison, James,   -1899, colliery manager.
Eng - 88, 336.

Harrison, James William, 18(17)-97, manufacturer.
Eng - 83, 253.

Harrison, John, 1828-94, surveyor.
PICE - 118, 447-448.

Harrison, John Atkinson, 1816-79, railway engineer.
PICE - 60, 407-408.

Harrison, John Thornhill, 1815-91, inspecting local govt engr.
Engng - 52, 574.

        Iron - 38, 430.
        PICE - 109, 405-407.
        PIMCE - 18, 439-440.

Harrison, Joseph, 1826-99, railway engineer overseas.
        PICE - 136, 346-347.

Harrison, Robert John, 1848-97, municipal engineer.
        Engng - 64, 572.
        PICE - 131, 384-385.

Harrison, Thomas Elliott, 1808-88, chief railway engineer(B,D).
        Eng - 65, 234, 263.
        Engng - 45, 294, 310, 359. Port. 357.
        Iron - 31, 277.
        PICE - 94, 301-313+frontis port.
        PIME - 1888, 261-263.

Harrison, William Arthur, 18(30)-84, steel manufacturer.
        PIME - 1884, 402.

Harrowing, Robert, 18(25)-1900, shipowner.
        Eng - 90, 304.

Hart, H   C   , 18(46)-95, telegraph engineer.
        Elec - 35, 599.
        Elec Rev - 37, 296.

Hart, Henry Sebastian, 1868-99, civil engineer.
        PICE - 137, 441.

Hartley, Frederick William, 1829-86, gas engineer.
        PICE - 87, 450-451.

Hartley, James, 1810-86, glass manufacturer(B).
        PICE - 85, 409-412.

Hartley, Jesse, 1780-1860, dock engineer of Liverpool(B,D).
        PICE - 33, 219-223.

Hartley, John,   -1884, ironmaster.
        JISI - 25, 560-561.
        PICE - 85, 409.

Hartley, John Bernard, 1814-69, dock engineer.
        PICE - 32, 216-219, 222-223.

Hartley, John Edward, 1831-83, railway engineer overseas.
        PICE - 76, 359-360.

Hartree, William, 1813-59, mechanical engineer.
        PICE - 19, 174-175.

Hartwright, John Henry,   -1869, water engineer.
        PICE - 30, 437.

Harvey, Charles Randolph, 1846-99, machine tool manufacturer.
Eng - 88, 191.
PIME - 1899, 616.
TIESS - 42, 408.

Harvey, Edward Cartwright, 1866-94, mining engr in S.Africa.
PIME - 1885, 311-312.
TSE - 1894, 257.

Harvey, Henry Nicholas, 1857-92, engineering manufacturer.
PICE - 112, 368.

Harvey, N   , 18(57)-92, shipbuilder & mining engineer.
Iron - 40, 560.

Harvey, Nicholas Oliver, 1801-61, mechanical engineer.
PICE - 21, 558-560.

Harvey, Robert, 1812-83, boiler maker.
Engng - 35, 438-439.

Harwood, Francis,   -1896, telegraph engineer in Spain.
Elec Rev - 39, 805.

Harwood, Hamilton Edward, 1824-72, architect & surveyor.
PICE - 38, 313-314.

Haskins, William, 1828-96, municipal engineer in Canada.
PICE - 126, 399.

Haslam, George, 1816-78, railway engineer.
TCMCIE - 7, 9-10.

Hastie, John, 18(44)-94, marine engineering manufacturer.
Eng - 78, 303.
Engng - 58, 422.
TIESS - 38, 334.

Haswell, John, 1812-97, mechanical engineer in Europe.
Eng - 84, 31-32+port.
TSE - 1897, 203.

Haughton, Samuel Wilfred, 1822-99, railway engineer.
PIME - 1899, 616-617.

Hawkes, Edward Claude, 1845-90, Public Works engr in India.
PICE - 101, 308-309.

Hawkes, Francis, 1818-80, railway engineer in India.
PICE - 68, 295-296.

Hawkes, William, 1800-62, engineering manufacturer.
PIME - 1863, 13.

Hawkins, Charles William, 1839-75, railway engr in India.
    PICE - 42, 258-259.
    PIME - 1876, 22.

Hawkins, Edward Jackson, 1840-90, surveyor.
    PICE - 109, 420-421.

Hawkins, George, 1815-88, railway manager.
    PICE - 95, 397-398.

Hawkins, John Isaac, 1772-1865, inventor & consulting engr(B).
    PICE - 25, 512-514.

Hawkshaw, Sir John, F.R.S., 1811-91, consulting civil engr(B,D).
    Engng - 51, 679.
    Iron - 37, 513.
    PICE - 106, 321-335 +port. frontis.
    Prac Mag - 7, 65-68.

Hawksley, Thomas, F.R.S., 1807-93, consulting civil engineer(B,D).
    Elec - 30, 574*.
    Eng - 76, 311.
    Engng - 56, 395-396.
    JISI - 44, 290.
    PICE - 117, 364-376.

Hawley, John Prentis, 1818-85, waterworks engineer.
    PICE - 80, 34?-343.

Hawthorn, Robert, 1796-1867, mechanical engineer.
    PICE - 27, 590-592.
    PIME - 1868, 15.

Hawthorn,Thomas, 1838-80, locomotive manufacturer.
    Eng - 50, 164*.
    JISI - 16, 691-692.
    PIME - 1881, 4-5.

Hayes, Edward, 1818-77, engine manufacturer.
    Eng - 44, 183.

Hayes, John, 1863-95, draughtsman.
    PIME - 1896, 94-95.

Hayes, Richard Frederic Fitz-Edmund, 1855-95, civil engineer.
    PICE - 124, 427-428.

Haynes, Capt.Henry Sidney Freeman, 1845-80, Royal Engineer.
    PICE - 63, 327.

Haynes, Thomas John, 1837-87, engineer in Gibraltar.
    PIME - 1887, 147.

Hayter, Harrison, 1825-98, consulting civil engineer(B).
    Eng - 85, 452.
    Engng - 65, 604, 627.
    PICE - 134, 391-394.
    PIME - 1898, 533-534.

Hayward, William, 1841-90, works manager.
    PICE - 103, 387.

Haywood, William, 1821-94, municipal engineer(B,D).
    Elec - 32, 683.
    Eng - 77, 335.
    PICE - 117, 376-378.
    PIMCE - 20, 378-379.

Hazeldine, William, 1763-1840, iron founder(D).
    CEAJ - 4, 48-49.

Head, Jeremiah, 1835-99, consulting mechanical engineer.
    Elec Rev - 44, 473.
    Eng - 87, 259, 378*.
    Engng - 67, 354-355+port.
    JISI - 55, 263-264.
    PICE - 136, 347-350.
    PIME - 1899, 134-137.

Head, John, 1832-81, pumping engine manufacturer(B).
    Eng - 51, 413.
    Engng - 31, 547-548.
    Iron - 17, 379.
    JISI - 19, 574-575.
    PICE - 67, 397-399.
    PIME - 1882, 8-9.

Head, John, 1839-93, mechanical engineer.
    Eng - 76, 65.
    Engng - 56, 79.
    JISI - 43, 174-176.
    PICE - 114, 378-380.

Head, Thomas Howard, 1833-80, consulting engineer.
    PICE - 61, 297-298.

Hearne, Edward Beresford, 1854-96, railway engineer.
    PICE - 126, 402-403.

Heath, Robert, 1816-93, ironmaster(B).
    JISI - 44, 290-291.

Heath, William Jerry Walker, 1835-69, Public Works engr in India.
    PICE - 30, 472.
    PIME - 1870, 14.

Hedley, Edward, 1836-82, consulting mining engineer.
   PICE - 72, 314-315.
   TCMCIE - 12, 17-18.

Hedley, John, 1817-64, mining engineer.
   PIME - 1865, 15.

Heinke, John William, 1816-70, submarine engineer.
   Engng - 9, 295*.

Helson, J   B   ,   -1879, forge manager.
   JISI - 15, 613*.

Hemans, George Willoughby, 1814-85, railway engineer(B).
   PICE - 85, 394-399.

Hemberow, Robert Cadding, 1849-88, railway engineer overseas.
   PICE - 92, 395-396.

Henderson, Andrew, 1800-68, steamship captain (B).
   PICE - 30, 472-475.

Henderson, David, 18(17)-93, shipbuilder.
   Engng - 56, 790.
   TIESS - 37, 195-196.

Henderson, Lt.-Col. George, 1783-1855, Royal Engineer(B,D).
   PICE - 15, 100-101.

Henderson, John, 18(18)-97, iron & steel manufacturer.
   Eng - 83, 401.

Henderson, John, 18(46)-1900, shipbuilder.
   TIESS - 44, 342.

Henderson, John M   , 18(42)-1900, machine tool manufacturer.
   Eng - 90, 440.

Henderson, Peter Lindsay, 1831-81, ship owner(B).
   PICE - 64, 341-342.

Henderson, Thomas, 18(20)-95, ship owner(B).
   Eng - 79, 174.

Henderson, Thomas, 1843-96, mechanical engineer.
   PIME - 1896, 258.

Henderson, William, 1825-95, shipbuilder.
   Engng - 59, 474.
   TIESS - 39, 270.

Henderson, William, 1827-81, metallurgist.
   Engng - 31, 90-91.
   JISI - 19, 573.

Henderson, William, 1854-98, consulting civil engineer.
PICE - 136, 350-351.

Henderson, William Scott, 1844-81, railway engineer.
PICE - 71, 421-422.

Hendry, George T   , 1824-96, ---
TIESS - 40, 257*.

Henesey, Richard, 1837-92, iron works engineer.
PIME - 1893, 494.

Henley, George,   -1888, telegraph engineer.
Elec - 20, 227*.
Elec Rev - 23, 36*.

Henley, William Thomas, 1814-82, telegraph engineer(B,D).
Elec - 10, 136.
Eng - 54, 471.
Iron - 20, 530.

Hennet, George, 1799-1857, railway contractor.
PICE - 17, 100-102.

Henshaw, Alfred,   -1900, railway engineer.
Eng - 89, 394.

Hensman, Henry, 1813-97, consulting civil engineer.
PICE - 131, 368.

Hepburn, Alexander, 1848-89, consulting marine engineer.
NECIES - 6, lxii-lxiii.

Heppel, John Mortimer, 1817-72, railway engineer overseas(B).
PICE - 36, 265-268.

Heppell, Thomas,   -1898, mining engineer.
Eng - 86, 460.
TIME - 21, 6.

Herapath, Spencer, 1822-84, railway director & broker(B).
PICE - 78, 447-448.

Herbert, Charles, 18(54)-98, works manager.
JISI - 53, 316.

Heriot, Gerald Maitland, 1868-1900, railway engineer.
PICE - 143, 333.

Hetherington, John, 1798-1864, iron founder.
Eng - 16, 376*.

Hetherington, Thomas Ridley, 1835-89, machine tool manufr.
PICE - 101, 295.

Hewett, Daniel Pinkney, 1818-54, civil engineer & architect.
PICE - 14, 136.

Hewett, Edward Edwards, 1843-84, consulting engineer.
PIME - 1884, 473.

Hewitson, William Watson, 1815-63, locomotive manufr.
PIME - 1864, 17.

Hewitt, John Richardson, 1842-93, consulting civil & mining
engr.
PICE - 113, 349-350.

Hey, William, 1852-86, mining engineer.
TCMCIE - 16, 13.

Hick, Benjamin, 1790-1842, mechanical engineer.
PICE - 2, 12-13.

Hick, Benjamin, 18(45)-82, engineering manufacturer.
Eng - 54, 272.
Engng - 34, 341.

Hick, John, 1815-94, engineering manufacturer(B).
Eng - 77, 120+port.
Engng - 57, 199.
JISI - 45, 391.
PICE - 117, 379-380.
PIME - 1894, 161-162.
TMAE - 4, 227-228.

Hickson, Sir Joseph, 1830-97, railway engineer(B).
Eng - 83, 57.

Hide, Thomas Comings, 1825-91, consulting marine engineer.
PIME - 1891, 606-608.

Higgin, George, 1833-92, civil engineer overseas(B).
PICE - 112, 349-353.

Higginson, Harry Pasley, 1838-1900, civil engineer overseas.
PICE - 142, 357-358.

Higginson, Thomas,   -1895, municipal surveyor.
PIMCE - 24, 368*.

Highton, Edward, 1817-59, telegraph engineer(B).
PICE - 19, 188-189.

Higinbotham, Thomas, 1820-80, rly engineer in Victoria.
Eng - 50, 421.
PICE - 63, 314-316.

Higson, Peter, 18(38)-80, mining engineer.
   Eng - 50, 398.

Hildred, Jesse, 1843-75, mining surveyor.
   PICE - 44, 228-229.

Hill, Alfred, 18(54)-1900, manufacturer.
   JISI - 57, 253*.

Hill, Alfred C   , 1835-89, iron & steel manufacturer.
   JISI - 36, 177.
   PIME - 1890, 172-173.

Hill, Francis,   -1894, ironworks manager.
   JISI - 45, 391-392.

Hill, John, 1812-94, surveyor.
   PICE - 116, 361-362.
   PIMCE - 20, 381-382.

Hill, Lawrence, 1816-92, consulting marine engineer.
   TIESS - 36, 319-320.

Hills, Staff-Cmdr Graham Hewett, R.N., 1826-88, marine engineer.
   PICE - 95, 398-405.

Hindle, Joseph, 1838-92, rly engineer in South America.
   PICE - 111, 378-380.

Hindmarsh, Thomas, 1827-89, railway engr in India.
   PICE - 97, 402-403.
   PIME - 1889, 193-194.

Hinds, Herbert, 1864-96, civil engineer.
   PICE - 126, 403.

Hindson, William, 1846-96, design engineer.
   NECIES - 12, 254.
   PIME - 1896, 599.

Hingeston, Charles Hilton, 1858-90, engineer in Gambia.
   PICE - 104, 311-312.
   TSE - 1890, 214.

Hingley, Noah,   -1877, coal & ironmaster.
   JISI - 11, 540*.

Hirst, T   , 18(24)-1900, iron works manager.
   Eng - 89, 318*.

Hobbs, William Fisher, 1809-66, agricultural engineer(B).
   PICE - 26, 577-579.

Hobson, Anthony Grant, 1861-87, railway surveyor overseas.
   PICE - 91, 449-450.

Hocking, Samuel, 1807-77, mining engineer.
    PICE - 51, 274.

Hodge, Robert, 1810-96, civil & mining engineer.
    PIMCE - 17, 288-289.

Hodges, Edward, 1844-92, Public Works engr in India.
    PICE - 113, 352-353.

Hodges, James, 1814-79, railway contractor(B).
    Engng - 28, 78-79.

Hodgetts, Alfred, 1830-87, coal & ironmaster.
    JISI - 32, 218.

Hodgkin, Maj. Elliot, 18(56)-1900, ironmaster.
    Elec Rev - 46, 191.
    Eng - 89, 106.
    Engng - 69, 123.

Hodgkinson, Eaton, F.R.S., 1789-1861, Prof. of Engineering,
    University College, London(B,D).
    PICE - 21, 542-545.
    PIME - 1862, 14-17.

Hodgson, John, 1814-57, consulting engr & manufacturer.
    PICE - 18, 204-205.

Hodgson, John,   -1895, iron merchant.
    JISI - 47, 261*.

Hodgson, John Lee, 18(25)-1900, mechanical engineer.
    JISI - 58, 389.

Hodgson, Robert, 1817-77, railway superintendent.
    PICE - 50, 178-181.
    PIME - 1878, 10.

Hodson, Richard, 1831-90, design engineer.
    PICE - 106, 335-336.
    PIME - 1890, 556.

Hoey, D   G   ,   -1891, ventilating engineer.
    Iron - 38, 541.

Hoey, Thomas,   -1874, shipbuilder.
    Engng - 17, 341.

Hoffman, Albert,   -1897, iron merchant.
    Eng - 84, 311.

Hogg, Thomas, 1852-94, surveyor.
    TSE - 1894, 253-254.

Holden, Sir Isaac, Bart., 1807-97, inventor; manufr(B).
Eng - 84, 178+port.
Engng - 64, 237-238.

Holford, James, 1820-82, colliery owner.
TCMCIE - 11, 16.

Holliday, John, 1825-85, tar works manager.
PIME - 1885, 161.

Holliday, Thomas, 18(41)-98, electrician; chemist.
Elec Rev - 42, 302.

Hollingworth, Andrew, 1837-84, engineer in Sweden.
Iron - 24, 302.

Holmes, James Archibald Hamilton, 1836-72, engineer in India.
PICE - 39, 290-291.

Holmes, Nathaniel John,   -1888, pioneer telegraph engineer.
Elec Rev - 23, 294.

Holmes, Philip Harrison, 1854-81, civil engineer.
PICE - 70, 432-433.

Holmes, Samuel Furness, 1821-82, municipal surveyor.
PICE - 72, 315.

Holmes, Sheriton, 18(29)-1900, civil engineer.
Eng - 89, 487*, 488.

Holmes, William, 1823-87, locomotive engineer.
Engng - 44, 391.

Holst, L   M   , 18(50)-89, telegraph engineer in China.
Elec Rev - 25, 614.

Holt, Arthur, 18(51)-94, municipal engineer.
PIMCE - 20, 382.

Holt, Francis, 1825-93, locomotive works engineer.
PIME - 1893, 91.
TIME - 5, 481.

Holt, William Lyster, 1839-89, tramway engineer.
PIME - 1889, 194-195.

Holtzapffel, Charles, 1806-47, mechanical engineer.
CEAJ - 10, 163.
PICE - 7, 14-15.

Holtzapffel, John Jacob, 1836-97, mechanical engr & writer.
PICE - 131, 385-387.

Homer, Charles James, 1837-93,coal & ironmaster.
    JISI - 45, 392*.
    PICE - 116, 362-364.
    PIME - 1893, 495-496.

Homersham, Samuel Collett, 1855-92, consulting water engr.
    PICE - 109, 421-422.

Hoof, James, 1821-49, railway contractor.
    PICE - 10, 97.

Hooper, Edward, 1822-69, railway engineer.
    PICE - 31, 248-249.

Hooper, Henry, 1828-82, Public Works engr in India.
    PICE - 71, 399-400.

Hopkins, James Innes, 1837-74, ironmaster.
    Eng - 37, 364*.
    Engng - 17, 408*.
    PICE - 39, 291.
    PIME - 1875, 26.

Hopkins, John Satchell, 1825-98, tinplate manufacturer(B).
    PIME - 1898, 138.

Hopkins, Rice, 1807-57, consultant engineer.
    PICE - 18, 192.

Hopkins, Thomas, 1809-48, ironmaster.
    PICE - 8, 11-12.

Hopkinson, Charles Herbert, 1858-92, railway engr overseas.
    PICE - 112, 369.

Hopkinson, Dr John, F.R.S., 1849-98, electrical engineer(D).
    Elec - 41, 622-623, 641.
    Elec Rev - 43, 338-339+port, 386, 463, 557.
    Eng - 86, 234+port, 506.
    Engng - 66, 301-303+port, 324, 531, 651, 783.
    JIEE - 28, 676-678.
    PICE - 135, 338-349.
    PIME - 1898, 534-536.

Hopper, William, 1816-85, foundry owner in Russia.
    PIME - 1886, 121.

Horley, Edward,   -1899, colliery owner.
    Eng - 88, 179.

Horn, George William, 1801-73, railway auditor.
    PICE - 38, 314-315.

Horn, Thomas William, 1864-97, railway engineer overseas.
    PICE - 129, 394.

Horne, Arnold, 1855-83, Public Works engineer in India.
PICE - 75, 314-315.

Horne, James, F.R.S., 1790-1856, amateur engineer(B).
PICE - 17, 102-103.

Horsley, Charles Cressy, 1856-94, consulting engineer.
PICE - 118, 460.
TSE - 1894, 255-256.

Horsley, Thomas, 1825-85, ironworks manager.
PIME - 1885, 525-526.

Hosking, John, 1808-71, ironworks engineer.
PIME - 1872, 17-19.

Hosking, John, 1839-74, ironworks engineer.
PIME - 1875, 26.

Hosking, Richard, 1819-82, iron mine manager.
Iron - 19, 307.
JISI - 23, 668-669.

Houghton, Francis Gassiot, 1858-97, consulting mech. engr.
PIME - 1897, 234-235.

Houghton, George, 1841-70, mechanical engineer.
PICE - 31, 249-250.

Houghton, William,   -1893, ironmaster.
Iron - 41, 385.

Houldsworth, John, 1807-59, ironmaster(B).
PICE - 19, 189-190.

Hovenden, <u>Brevet Lt.-Col.</u> Julian St.John, 1831-70, Public Works
engr in India.
PICE - 31, 250-252.

Howard, James, 1821-89, agricultural implement manufr(B).
Eng - 67, 90.
Iron - 33, 102.
JISI - 34, 218.

Howard, Thomas, 1796-1872, ironmaster.
PICE - 36, 294-296.

Howard, Thomas, 1816-96, dock engineer, Bristol.
Eng - 81, 122.
Engng - 61, 164.
PICE - 124, 413-414.

Howard, William Frederick, 1829-99, mining engineer.
TFIME - 18, 151-152.

Howard-Keeling, Herbert, 1829-93, ironmaster.
    PICE - 114, 396-397.
    PIME - 1893, 388.

Howden, Andrew Cassels, 1837-75, engineer in India.
    PICE - 42, 263.

Howe, Elias, 1819-67, inventor of sewing machine.
    Eng - 24, 337*.
    Prac Mag - 5, 321-324.

Howe, William, 1814-79, colliery engineer(B).
    Eng - 47, 67.
    PIME - 1880, 6-8.
    TCMCIE - 7, 10-12.

Hownam-Meek, Sydney Hownam, 1854-98, railway engineer.
    Eng - 85, 118.
    PICE - 132, 382-383.

Hudson, Frederick William, 1859-98, dock engineer.
    Engng - 65, 369.
    PICE - 132, 395.

Hudson, Henry Walter, 1841-96, rly engineer in India.
    PICE - 131, 368-369.

Hudson, Dr R   S   ,   -1883, Secy, Miners' Association
    of Cornwall and Devon.
    Iron - 22, 467.

Huffman, Solomon, 1854-84, metallurgist.
    Engng - 39, 34*.

Hughes, John, 1814-89, ironmaster in Russia.
    Eng - 68, 38.
    Iron - 34, 31.
    JISI - 36, 177-178.

Hughes, John d'Urban, 1807-74, civil engineer overseas.
    PICE - 40, 255-258.

Hughes, Robert, 1800-60, admiralty engineer.
    Eng - 10, 242.
    PICE - 20, 163-164.

Hughes, Samuel,   -1870, gas engineer.
    Eng - 30, 358*.
    Engng - 10, 393*.

Hughes-Hallett, Charles Frederick, 18(53)-1900, mechanical engr.
    Eng - 89, 673.
    PICE - 142, 387.

Huish, Capt. Mark, 1808-67, railway manager(B).
PICE - 27, 600-602.

Hulbert, Henry George, 1839-71, surveyor.
PICE - 33, 265.

Hulse, Joseph Whitworth, 1861-98, engineering manufacturer.
Eng - 85, 465.
Engng - 65, 565*.
JISI - 53, 316-317.
PIME - 1898, 314.
TMAE - 8, 220.

Hulse, William Wilson, 1821-97, mechanical engineer.
Eng - 83, 326+port.
Engng - 63, 414.
PICE - 128, 347-348.
PIME - 1897, 137.

Humber, William, 1821-81, consulting civil engineer(B).
PICE - 65, 375-377.

Humble, Edward Barber 1841-73, Bengal Engineer.
PICE - 41, 226-227.

Humble, Joseph, 1843-87, colliery manager.
TCMCIE - 17, 18-19.

Humphreys, Henry Temple, 1839-91, Public Works engr in India.
PICE - 108, 403-404.

Humphrys, Edward, 1808-67, marine engineer.
Eng - 23, 478; 87, 82-83+port.
PICE - 27, 592-595.
PIME - 1868, 15-16.

Hunt, Sir Henry Arthur, 1810-89, consulting surveyor(B).
PICE - 99, 398-400.

Hunt, Herbert Edgell, 1847-96, railway engr overseas.
PICE - 126, 404-405.

Hunt, Robert, F.R.S., 1807-87, mining engineer(B,D).
Eng - 64, 335.
Iron - 30, 377.

Hunt, Thomas, 18(16)-96, railway engineer.
PIME - 1898, 536-537.

Hunt, William, 1843-97, chief railway engineer.
Eng - 83, 373.
Engng - 63, 448-449.
PICE - 129, 372-374.

Hunter, James, 1818-86, ironmaster(B).
    Engng - 42, 369*.
    Iron - 28, 338.
    JISI - 29, 800-801.
    PICE - 89, 494-495.
    TIESS - 30, 311-312.

Hunter, James Bernard, 1855-99, engineering manufacturer.
    Eng - 87, 434.
    Engng - 67, 549*, 579.
    PICE - 136, 351.

Hunter, John, -18(85), ironmaster.
    TIESS - 29, 212-220.

Hunter, Michael, -1898, tool manufacturer.
    Eng - 86, 605.

Hunter, Robert, 1840-98, gas works manager.
    TSE - 1898, 239.

Hunter, Walter, 1772-1852, mechanical engineer(B).
    PICE - 12, 161-163.

Hunter, William Havard, 1835-91, surveyor.
    PICE - 109, 429.

Huntley, William, 1798-1880, railway engineer.
    Engng - 29, 314.
    Iron - 15, 299.

Hunton, H , -1893, telegraph engineer.
    Elec - 31, 166.

Hurst, Thomas Grainge, 1824-90, mining engineer.
    PICE - 103, 366-368.

Hurwood, George, -1864, dock engineer, Ipswich.
    PICE - 24, 531-532.

Husband, William, 1822-87, hydraulic & mining engineer(B).
    Eng - 63, 361.
    Iron - 29, 384.
    PICE - 89, 470-473.

Hutchings, Henry Burdon, 1854-80, Public Works engr in India.
    PICE - 64, 341.

Hutchinson, George, -1890, draughtsman.
    Eng - 70, 481.

Hutchinson, Henry, 18(13)-98, surgical instrument manufr.
    Engng - 65, 112.

Hutton, Robert,   -1899, foundryman.
    JISI - 55, 265*.

Hutton, Robert Joseph, 1842-90, railway engineer.
    PICE - 104, 293-294.
    TSE - 1892, 243.

Huxham, Hortensius, 1848-1900, mining engineer.
    Eng - 89, 194.
    PICE - 140, 273-274.

Huxley, Herbert George, 1847-93, water engineer overseas.
    PICE - 114, 388-389.

Hyde, Maj.-Gen. Henry, 1825-87, Bengal Engineer(B).
    Iron - 30, 353.
    JISI - 28, 217-218.
    PICE - 91, 462-466.
    PIME - 1887, 467-468.

Hyde, Mark, 1823-93, railway engineer.
    PICE - 117, 380-381.

Hynes, Frederick Margarson, 1842-92, Public Works engineer in
    Victoria.
    PICE - 114, 380-381.

Hyslop, Peter Sharp, 1852-99, civil engineer.
    PICE - 137, 428-429.

I'Anson, Charles,   -1884, ironmaster.
    Iron - 24, 251.
    JISI - 25, 559-560.

I'Anson, James, 1845-98, foundry owner.
    Eng - 85, 333.
    JISI - 53, 317.

Inglis, Anthony, 1813-84, shipbuilder & engineer.
    Engng - 37, 110.
    TIESS - 27, 214.

Inglis, J   Anthony, 1862-96,   foundry owner.
    TIESS - 39, 270-271.

Inglis, John, 18(30)-83, consulting engr in Hong Kong.
PIME - 1884, 64.

Inglis, John, -1888, shipbuilder & marine engr.
Engng - 45, 495-496; 46, 332*.
Iron - 31, 440.
TIESS - 31, 236.

Inglis, William, 1835-90, engine manufacturer.
Eng - 69, 339.
Engng - 49, 514.
PICE - 101, 296-297.

Inman, Thomas, 1848-98, Public Works engr in India.
PICE - 132, 395.
TSE - 1898, 233.

Inman, William, 1825-81, ship owner.
Eng - 52, 31.
Iron - 18, 56.

Innes, , -1898, railway engineer overseas.
Eng - 86, 519*.

Innes, Cosmo, 1841-87, railway engineer in India.
PICE - 90, 433-434.

Innes, Edward Arthur Robert, 1852-87, harbour engr overseas.
PICE - 92, 396-397.

Innes, Capt. William, 1841-71, Royal Engineer.
PICE - 43, 312-317.

Inshaw, John, 1807-93, mechanical engineer(B).
Engng - 55, 70.
Iron - 41, 53.

Insole, J W , -1898, colliery owner.
Eng - 85, 466*.

Irvine, Col. Archibald, 1797-1849, Royal Engineer.
PICE - 10, 87-90.

Irwin, Thomas Frederick, 1848-97, naval architect.
NECIES - 13, 270-271.
PIME - 1897, 235.

Ismay, Thomas Henry, 1837-99, shipowner(B).
Eng - 88, 549.
Engng - 68, 693.
TINA - 42, 297.

Jack, Alexander,   -1877, agricultural implement manuf-
    acturer.
    Engng - 23, 447.

Jack, Alexander, 1846-86, engineering manufacturer.
    Eng - 61, 237*.
    PIME - 1886, 462-463.

Jackson, Edward Patten, 1842-81, colliery owner.
    TCMCIE - 9, 100.

Jackson, Edward Rainford, 1860-89, engineer in Jamaica.
    PICE - 100, 412.

Jackson, Edward Wilthew, 1838-95, railway engineer overseas.
    PICE - 124, 428-429.

Jackson, Henry James, 1824-84, shipping engineer(B).
    PICE - 80, 332-333,
    PIME - 1884, 473-474.

Jackson, John, 1834-91, railway contractor overseas.
    PICE - 105, 324-325.

Jackson, John Peter, 1843-99, coal & ironmaster.
    Eng - 87, 248.
    JISI - 55, 265.
    PICE - 136, 351.
    TFIME - 18, 150-151.

Jackson, Joseph,   -1860, railway contractor.
    Eng - 10, 205.

Jackson, Peter Rothwell, 1813-99, inventor & manufacturer.
    PIME - 1899, 137-139.

Jackson, Ralph Ward, 1806-80, developer of West Hartlepool(B).
    PICE - 63, 328-332.

Jackson, Robert,   -1873, implement manufacturer.
    Engng - 16, 63*.

Jackson, Col. Robert Raynsford, 1823-98, telegraph company
    chairman.
    Elec Rev - 43, 21.
    JIEE - 28, 678.

Jackson, Thomas, 1808-85, railway contractor(B).
    Eng - 59, 50.
    Engng - 39, 59.
    Iron - 25, 53.

Jackson, Sir William, Bart., 1805-76, contractor(B).
PICE - 45, 252-256.

Jackson, William,   -1885, mechanical engineer.
PICE - 83, 441-442.

Jacob, Arthur, 1831-95, municipal engineer.
Eng - 79, 143-144.
PICE - 120, 355-356.

Jacob, Charles Ryder, 1867-97, naval engineer.
PICE - 131, 387.

Jacomb, William, 1832-87, chief railway engineer(B).
Eng - 63, 437.
Iron - 29, 490.
PICE - 90, 434-435.

Jacomb-Hood, Robert, 1822-1900, railway engr & contractor(B).
PICE - 142, 359-361.

James, Arthur, 1816-66, railway engineer.
PICE - 26, 580-582.

James, Christopher, 1836-95, consulting engr & inventor.
PIME - 1895, 143-144.

James, D   W   ,   -1897, colliery agent.
Eng - 83, 652*.

James, David,   -1900, Secretary, Dowlais Iron Company.
Eng - 89, 266.

James, Maj.-Gen. Sir Henry, F.R.S., 1803-77, Royal Engr(B,D).
Eng - 44, 2.

James, Jabez, 1810-83, machinery manufacturer(B).
PICE - 73, 358-360.
PIME - 1884, 64-65.

James, James,   -1892, furnace manager.
Iron - 40, 296*.

James, John, 1825-73, iron & tinplate manufacturer.
PIME - 1874, 21-22.

James, Thomas, 18(22)-89, ironworks engineer.
Iron - 33, 185.
JISI - 34, 219-220.

James, William, 1854-89, contractor in India.
PICE - 99, 376-377.

Jameson, James Wardrop, 1824-60, marine engineer.
PICE - 20, 164-166.

Jamieson, John Lennox Kincaid, 1826-83, shipbuilder(B).
Engng - 36, 18, 88.
Iron - 22, 10.
PIME - 1884, 65-66.
TIESS - 27, 214.

Jamieson, Mathew Buchan, 1860-95, civil engineer overseas.
Engng - 60, 271.
PICE - 123, 442-444.

Janson, Ailsa, 1844-85, railway engineer overseas(B).
PICE - 81, 324-327.

Jee, Alfred Stanistree, 1816-58, railway engineer.
PICE - 18, 193-196.

Jeffcock, Parkin, 1829-66, mining engineer(B).
Eng - 22, 483.
PICE - 27, 594-595.
PIME - 1867, 15-16.

Jeffcock, Thomas William, 18(40)-1900, civil & mining engr.
Eng - 90, 136, 278.

Jefferies, John Robert, 1840-1900, steam plough manufr.
Elec Rev - 47, 473*.
Engng - 70, 342*, 370.
JISI - 58, 389-390.
PIME - 1900, 626-627.

Jeffrey, Robert, 1813-77, railway engineer overseas(B).
PICE - 52, 273-276.

Jeffreys, Edward Alexander, 1824-89, ironmaster.
JISI - 34, 220-221.
PICE - 96, 319-320.
PIME - 1889, 334-335.

Jenkin, Henry Charles Fleeming,F.R.S., 1833-85, Professor of
Civil & Mechanical Engineering, Edinburgh University(B,D).
Elec - 15, 97.
Eng - 59, 483.
Engng - 39, 680-681.
Iron - 25, 56.
JIEE - 14, 345.
PICE - 82, 365-377.
PIME - 1885, 458.

Jenkins, Charles Morris, 1866-99, railway engineer in Natal.
Eng - 88, 632.
PICE - 141, 351-352.

Jenkins, E   , 1834-88, manufacturing engineer.
   Engng - 46, 404.

Jenkins, Philip, 1854-91, Prof. of Naval Architecture,
   Glasgow University(B).
   Eng - 71, 491.
   Engng - 51, 739.
   Iron - 37, 537.
   TIESS - 34, 322-323.
   TINA - 33, 303-304.

Jenkins, Walter, 1868-1900, Public Works engr in India.
   PICE - 143, 333-334.

Jenkins, William, 1803-67, locomotive superintendent.
   PIME - 1868, 16.

Jenkins, William, 1825-95, ironworks manager.
   JISI - 47, 262.
   NECIES - 12, 246-247.

Jenkinson, Joseph William, 1855-85, hydraulic engineer.
   PICE - 80, 337-338.

Jenks, Isaac James, 18(45)-97, ironmaster.
   JISI - 51, 311.

Jenks, John,   -1892, ironmaster.
   Iron - 39, 185.

Jennings, James C   , 1827-94, iron ore merchant.
   JISI - 46, 264*.

Jervis, Col. George Ritso, 1794-1851, Bombay Engineer(B).
   PICE - 11, 105-109.

Jervois, Lt.-Gen. Sir William Francis Drummond, F.R.S., 1821-97,
   Royal Engineer(B,D).
   PICE - 130, 322-323.

Jesper, Charles, 1854-1900, railway manager.
   Eng - 5C, 519, 529.

Jessop, Joseph, 1825-83, crane manufacturer.
   PIME - 1884, 66.

Jessop, Thomas, 1804-87, iron & steel manufacturer(B).
   Engng - 44, 573*.

Jobson, Robert, 1817-72, railway engineer.
   PICE - 36, 296.

John, William, 1845-90, naval architect(B).
   Eng - 71, 8.

Engng - 51, 21.
Iron - 37, 18.
JISI - 39, 246-247.
IINA - 32, 251-252.

Johns, Jasper Wilson, 18(25)-91, railway engineer.
Iron - 38, 98.

Johnson, Alfred, 18(58)-1900, colliery owner.
Eng - 89, 657*.

Johnson, Cuthbert George Dixon, 18(42)-99, iron merchant.
Eng - 87, 199.

Johnson, John Clarke, 1840-1900, works manager.
PIME - 1900, 332.

Johnson, John Henry, 18(27)-1900, solicitor & patent agent.
Eng - 89, 336.
PICE - 141, 354-355.

Johnson, John Thewlis, 1836-96, wire manufacturer.
JISI - 49, 286.
PICE - 124, 438.

Johnson, John William, 1844-96, Public Works engr in India.
PICE - 127, 371.

Johnson, Thomas Marr, 1826-74, builder & contractor(B).
Engng - 18, 86.
PICE - 39, 268-269.

Johnson, Warwick Huson, 1846-99, Public Works engr in India.
PICE - 140, 274-275.

Johnson, William, 1823-64, patent agent(B).
PICE - 25, 528-529.

Johnson, William, 18(44)-1900, electrical engineer.
Elec Rev - 46, 925.
Eng - 89, 563.

Johnson, Lt.-Col. William Robert, 1830-82, Public Works engr
in India(B).
PICE - 72, 319-320.

Johnston, Andrew, 1818-84, chief railway engineer(B).
PICE - 78, 434-435.

Johnston, Andrew,   -1896, consulting marine engr in Hong Kong.
PIME - 1896, 95-96.
TIESS - 39, 271.

Johnston, Edward, 1816-80, surveyor in India.
PICE - 63, 319-320.

Johnston, John Montgomery, 1857-93, rly engineer in Brazil.
PICE - 116, 382-383.

Johnston, Thomas Masterman Hardy, 1817-94, rly engr.
PICE - 119, 393-394.

Johnston, W   P , -1889, telegraph engineer in India.
Elec - 23, 53*.

Johnstone, William, 1811-77, railway engineer(B).
Engng - 23, 346.
PICE - 49, 261-262.

Johnstone, William, 1844-79, railway engineer in India.
PICE - 62, 354-355.

Johnstone, Capt. William Henry, 1846-81, Royal Engineer.
PICE - 67, 415-417.

Joicey, John, M.P., 1816-81, iron & coal master(B).
JISI - 19, 577.
PICE - 69, 417-418.

Jones, Alfred Wilkinson, 1850-93, iron & steel manufacturer.
TSE - 1893, 237.

Jones, David, 18(44)-97, iron works manager.
JISI - 52, 255.

Jones, David Charles, 1867-94, sanitary engineer.
PICE - 118, 460-461.

Jones, Edward Francis,   -1892, ironmaster.
Iron - 40, 231.

Jones, Edward John, 1841-83, Public Works engr in India.
PICE - 75, 306-307.

Jones, Elias William, 1833-77, consulting engineer.
PICE - 50, 186-187.

Jones, Francis, 1815-97, Public Works engr in Ireland.
PICE - 129, 374-375.

Jones, Frederick,   -1874, railway engineer in Egypt.
Engng - 17, 283*.

Jones, Lt.-Gen. Sir Harry David, 1791-1866, Royal Engineer(B,D).
PICE - 30, 438-440.

Jones, Hodgson Monteith Layard, 1843-93, gas engineer in Europe.
PICE - 115, 400.

Jones, James, 1790-1864, gas works engineer.
PICE - 24, 532-533.

Jones, James, 18(36)-99, manufacturer.
JISI - 58, 390-391.

Jones, John, 1831-93, iron & steel works manager.
JISI - 44, 291*.

Jones, John, 1835-77, trade secretary(B D).
Eng - 43, 388*.
Engng - 23, 462.
JISI - 1877, App. C, viii-x.

Jones, John,   -1892, tinplate manufacturer.
Iron - 40, 362*.

Jones, John, 1842-96, railway secretary.
Eng - 81, 393*.
Engng - 61, 506.

Jones, John Hodgson, 1823-92, water & drainage engineer.
PICE - 112, 353-354.

Jones, Mark, 1804-63, mechanical engineer in India.
PICE - 23, 498*.

Jones, Rhys William, 1804-64, railway engineer.
PICE - 28, 608-610.

Jones, Richard Lionel, 1841-79, civil engineer.
PICE - 60, 404-405.

Jones, Robert, 1812-95, gas engineer.
PICE - 120, 356-357.

Jones, William George, 1858-92, railway engineer.
PICE - 109, 422- 423.

Jones, William Henry, 1853-87, railway engineer overseas.
PICE - 92, 397-398.

Jones, William John, 1849-84, gas & elec. engr in Italy.
PICE - 82, 385-386.

Jopling, Charles Michael, 1820-63, railway engineer.
PICE - 23, 508-511.

Jopling, Frederick, 1855-90, river engineer.
PICE - 101, 297-298.

Jopling, Thomas, 1842-94, steel works manager in USA.
JISI - 45, 392*.

Jopp, Charles, 18(20)-95, railway contractor.
Eng - 79, 326.
Engng - 59, 510.

Jordan, William, 1850-93, foundry manager in India.
  TSE - 1893, 238.

Jordon, Samson, 1831-1900, metallurgist.
  TSE - 1900, 267-268.

Jordon, Thomas Brown, 1807-90, consulting mining engr(B).
  Elec - 25, 165.
  Eng - 69, 495.
  Iron - 35, 541.

Joseph, Thomas, 1819-90, coal & ironmaster.
  PICE - 103, 368-370.

Josselyn, Frederick, 18(42)-1900, engineering manufacturer.
  Eng - 89, 52.
  Engng - 69, 62.

Jost, Adolf, 18(60)-98, steelworks manager.
  JISI - 55, 265-266*.

Jowitt, Albert A   , 1833-91, steel manufacturer.
  Iron - 38, 120.

Joyner, Henry Batson, 1839-84, civil engineer overseas.
  PICE - 79, 370-371.

Kane, John, 1819-76, Secretary, Ironworkers Association.
  Engng - 21, 238.

Kay, John Z   ,   -1878, ironmaster.
  Engng - 25, 319.

Keer, Thomas,   -1892, ironmaster.
  Iron - 40, 408.

Kelk, Sir John, Bart., 1816-86, contractor(B).
  PICE - 87, 451-455.

Kell, George Johnson, 18(38)-99, mining engineer.
  Engng - 67, 348.

Kelson, Frederick Colthurst, 1831-97, consulting marine engr.
  PIME - 1897, 137-138.
  TIESS - 41, 379.

Kemp, Dixon, 1839-99, naval architect(B).
    Eng - 88, 524.
    TINA - 42, 296.

Kemp, Ebenezer, 1831-92, boiler manufacturer.
    Eng - 74, 112*.
    Engng - 54, 208.
    TIESS - 35, 304-306.

Kemp, Henry, 1839-95, railway contractor overseas.
    PICE - 120, 357-358.

Kemp, John, 18(22)-92, agricultural implement manufacturer.
    Engng - 54, 263.
    Iron - 40, 184.

Kemp, William, 1847-97, draughtsman.
    TIESS - 40, 251-252.

Kempe, William,   -1883, mechanical engineer.
    Iron - 21, 78.

Kempe, William Henry Coryton, 1874-1900, railway engineer.
    PICE - 142, 387-388.

Kendall, Lt E   N   , R.N.,   -1845, ---
    PICE - 5, 5.

Kennard, Howard John, 1829-96, ironmaster.
    JISI - 50, 259-260.
    PICE - 126, 408-409.

Kennard, Robert William, 1800-70, ironmaster(B).
    Engng - 9, 29*.

Kennard, Thomas William, 1825-93, consulting civil engineer(B).
    Eng - 76, 269.
    Engng - 56, 330.

Kennedy, James, 1797-1886, engine manufacturer.
    Eng - 62, 268.
    Engng - 42, 351.
    Iron - 28, 305.
    PIME - 1886, 532-533.

Kennedy, Lt.-Col. John Pitt, 1796-1879, Royal Engineer(B,D).
    PICE - 59, 293-298.

Kennedy, John Pitt, 1824-97, railway engineer overseas.
    PICE - 131, 369-370.
    PIME - 1897, 515-516.

Kennedy, Myles, 1806-83, ironmaster.
    JISI - 23, 665.

Kennedy, Robert Baird, 1853-1900, mechanical engineer.
PIME - 1900, 627.

Kennedy, Thomas Stuart, 1844-94, mechanical engineer.
PICE - 120, 358-359.
PIME - 1894, 598.

Kennedy, William, 18(37)-99, oil engineer.
Engng - 67, 678.
JISI - 56, 294.

Kenrick, W W , -1892, mining engineer overseas.
Iron - 40, 319.

Kermode, Charles Chatterley, 18(47)-78, draughtsman.
PIME - 1879, 10.

Kerr, Andrew, -1888, municipal engr in Australia.
TSE - 1888, 253.

Kerr, Archibald, 18(44)-1900, pipe manufacturer.
TIESS - 43, 359-360.

Kerr, James, 1851-84, railway plant manufacturer.
PIME - 1885, 72-73.

Kerr, John, 1833-72,shipowner.
Engng - 13, 41*.

Kerr, Peter, 1818-69, thread manufacturer.
PIME - 1870, 14-15.

Kershaw, Burroughs Dickie, 1830-1900, water company director.
TSE - 1900, 268-269.

Kershaw, Durand, 1821-71, civil servant in Ceylon.
PICE - 36, 268-273.

Kershaw, Thomas, 1840-99, engineering lecturer.
PIME - 1899, 617-618.

Key, John, 18(20)-76, shipbuilder.
Engng - 21, 462.

Keydell, Amandus Edmund, 1841-92, Lloyds surveyor.
PIME - 1892, 404-405.

Kidd, Alexander, -1899, engineering surveyor.
TIESS - 43, 360.

Kilgour, Robert, 18(41)-95, mechanical engineer.
TMAE - 5, 265.

Kilvington, William, 1852-95, works manager.
    NECIES - 12, 247-248.
    PICE - 122, 372-373.

Kimpton, John George, 1839-87, consulting mining engr.
    TCMCIE - 17, 19-20.

Kincaid, Thomas, 1796-1884, marine engine manufacturer.
    Engng - 37, 275.

King, Alfred, 1797-1867, gas engineer(B).
    Mech Mag - 86, 278.
    PICE - 30, 440-441.

King, Frank, 18(54)-99, electrical engineer.
    Elec Rev - 45, 244.
    JIEE - 29, 955-956.

King, Henry J   H   , 18(45)-95, steam engine manufr.
    Engng - 60, 542.

King, Dr. J   T   , 18(34)-1900, patent lawyer.
    Eng - 90, 566.

King, Robert, 1822-91, gas engineer in Singapore.
    TSE - 1891, 215.

Kingdon, J   A   , 18(59)-96, electrical engineer.
    Elec Rev - 38, 603, 632.

Kinghorn, David, 1821-91, shipbuilder.
    Engng - 51, 565.
    Iron - 37, 425.
    TIESS - 34, 323.

Kinghorn, William, 1791-1874, contractor.
    Engng - 17, 341.

Kingston, William, 1780-1861, admiralty engineer.
    Eng - 12, 66*.

Kinsey, Alfred Burton, 1866-97, harbour engineer in Natal.
    TSE - 1897, 203.

Kircaldy, John,   -1887, manufacturer.
    Eng - 64, 150*.

Kirk, Alexander Carnegie, 1830-92, marine engr & shipbuilder(B).
    Eng - 74, 312.
    Engng - 54, 459, 487-488.
    Iron - 40, 319.
    PICE - 111, 380-383.
    PIME - 1892, 405-406.
    TIESS - 36, 320-323.
    TINA - 34, 238-239.

Kirkaldy, David, 1820-97, naval architect.
    Eng - 83, 118, 147+port.
    Engng - 63, 148-149.
    PICE - 128, 351-356.
    TIESS - 40, 252-255.

Kirkhouse, William, 18(14)-97, civil engineer.
    Engng - 64, 411.

Kirkwood, Thomas, 1850-84, dock & harbour engr overseas.
    PIME - 1885, 161.

Kirtley, Matthew, 1813-73, locomotive superintendent.
    PIME - 1874, 22.

Kitching, Alfred, 1808-82, iron founder(B).
    Engng - 33, 164*.
    JISI - 21, 658-659.

Kitson, Frederick William, 1829-77, rly plant manufacturer.
    JISI - 11, 538*.
    PICE - 52, 277.
    PIME - 1878, 11-12.

Kitson, James, 18(06)-85, locomotive engr & rly director(B).
    Engng - 40, 20.
    Iron - 26, 9.

Kitson, John Hawthorne, 1843-99, locomotive engineer.
    Eng - 87, 539.
    Engng - 67, 716.
    PICE - 137, 430-431.
    PIME - 1899, 269.

Kitson, William, 18(13)-75, locomotive superintendent.
    Engng - 20, 453.

Knight, John, 1850-99, municipal engineer.
    PICE - 139, 376-377.

Knight, John Peake, 1828-86, railway manager(B).
    Eng - 62, 95.
    PICE - 87, 456-458.

Knight, Samuel John, 1809-81, consulting civil engineer.
    PICE - 70, 434-435.

Knighton, Henry Amos, 1839-95, colliery manager.
    TFIME - 14, 173.

Knighton, Joseph Godber, 1851-95, foundry manager.
    PICE - 124, 430.

Knowles, John, 1828-94, colliery manager.
    PICE - 116, 364-365.

Knowles, Thomas, M.P., 1824-83, coal & ironmaster(B).
    Eng - 56, 453.

Kochs, W   E   , 18(34)-98, mining engineer.
    Engng - 65, 756.

Koe, Stephen Lancelot, 1827-99, ironworks manager.
    JISI - 55, 266.

Kohn, Frederick, 1837-71, metallurgist.
    PIME - 1872, 19.

Kunhardt, Maj. Henry Geffcken, 1850-92, Royal Engineer.
    Iron - 40, 427.
    PICE - 112, 376-378.

Kyd, Arthur Anderson, 1856-86, municipal engineer(B).
    PICE - 84, 449-450.

Laing, James, 1812-86, machinery manufacturer.
    Engng - 42, 291.

Laing, James, Jun.,   -1895, engineering manufacturer.
    NECIES - 12, 252-253.

Laing, Samuel, 1810-97, lawyer, railway chairman(B).
    Eng - 84, 157+port.

Laing, Thomas Elcoat, 1848-99, sanitary engineer.
    PICE - 140, 283.

Laing, William, 1837-99, marine superintendent.
    TIESS - 42, 408-409.

Laird, Henry Hyndman, 1838-93, shipbuilder.
    Eng - 75, 467.
    Engng - 55, 783.
    Iron - 41, 476.
    PIME - 1893, 205-206.
    TINA - 34, 239-240.

Laird, John, M.P., 1805-74, shipbuilder(B,D).
    Eng - 38, 343.
    Prac Mag - 3, 401-408.
    TINA - 16, 266-269+port.

Laird, John, 1834-98, shipbuilder(B).
    Eng - 85, 88, 138+port.
    Engng - 65, 121.

Laird, William, 1831-99,shipbuilder(B).
    Eng - 87, 134+port, 149*.
    Engng - 67, 188 +port.
    PIME - 1899, 139-140.
    TINA - 41, 374.

Lamb, Andrew,   -1881, marine superintendent.
    Engng - 31, 393.
    Iron - 17, 279*.

Lamb, Joseph Chatto, 18(03)-84, coal owner.
    Iron - 24, 451.

Lambert, Frank, 1836-87, telegraph engineer.
    Elec - 19, 415.

Lambert, George Nugent Reynolds, 1836-90, Public Works engr in
    India.
    PICE - 103, 375-376.

Lambert, William Blake, 1816-74, naval engr in Russia(B).
    PICE - 40, 255.
    PIME - 1875, 26-27.

Lambie, John, 18(34)-95, locomotive superintendent.
    Eng - 79, 113.
    Engng - 59, 178.

Lancaster, Charles William, 1820-78, armaments manufacturer(B,D).
    PICE - 53, 289-292.

Lancaster, John, 1815-84, iron & steel works chairman(B).
    JISI - 25, 558-559.
    PIME - 1884, 402.

Lancaster, Joshua,   -1899, coal & ironmaster.
    JISI - 55, 266.

Landale, David, 18(05)-95, mining engineer.
    Engng - 60, 800-801.

Lane, Christopher Bagot, 1814-77, consulting engr in Brazil(B).
    PICE - 48, 266-269.

Lane, Denny, 18(18)-95, gas & electrical engineer.
    Elec - 36, 177.

Lane, George Morgan, 1856-92, railway engineer overseas.
    PICE - 112, 370.

Lane, Michael, 1802-68, chief railway engineer.
    Eng - 25, 170.
    Mech Mag - 88, 186.
    PICE - 30, 441-442.

Lang, Edward Tickell, 1849-80, Public Works engr in India.
    PICE - 62, 355-356.

Lang, Oliver, -1867, naval architect.
    Mech Mag - 87, 111-112.

Lang, William, 1838-72, civil engineer.
    PICE - 36, 297-298.

Langdon, James Henry Cornwall, 1853-95, municipal engr in
    Australia.
    PICE - 124, 430-431.

Langé, Hermann Ludwig, 1837-92, engineering manufacturer.
    JISI - 42, 289-291.
    PICE - 109, 407-409.
    PIME - 1892, 406-407.

Langmuir, James, 1850-86, waterworks engineer.
    PICE - 88, 445-446.

Lanyon, Sir Charles, 1813-89, architect, civil engr(B,D).
    Eng - 67, 492.
    Iron - 33, 494.
    PICE - 98, 391-393.

Lanyon, John, 1839-1900, consulting civil engineer.
    PICE - 140, 275-276.

Larsen, Jorgen Daniel, 1833-91, tramway engineer.
    JISI - 39, 242-243.
    PICE - 104, 312-313.

Latham, Edwin Davenport, 1843-94, municipal surveyor.
    PICE - 120, 371-372.
    PIMCE - 21, 328.
    TSE - 1894, 257-258.

La Touche, Henry Christopher Digges, 1839-95, rly engr overseas.
    PICE - 120, 360-362.

Lavertine, Richard Aloysius, 1853-95, Bengal Engineer.
    PICE - 124, 431-432.

Law, David, 1816-69, ironmaster.
PIME - 1870, 15.

Law, Henry, 1824-1900, civil engineer.
Eng - 90, 321.
PICE - 142, 362-363.

Law, James,   -1892, ship owner.
Iron - 40, 78.

Lawrence, Frederick, 1828-64, ironmaster & civil engr.
PICE - 24, 541-542.

Laws, Hubert, 1841-91, railway contractor.
PICE - 97, 404-405.

Lawson, James Ibbs, 1861-96, rly engineer in New Zealand.
PICE- 127, 390.
PIME - 1896, 600.

Lawson, John, 1824-73, sanitary engineer.
PICE - 38, 315-317.

Lawson, John, 1837-95, contracting engineer.
PICE - 122, 373-374.

Laybourne, John, 1828-72, foundry owner.
PIME - 1873, 18.

Leach, Stephen William, 1818-81, Engineer, Port of London.
PICE - 70, 420-422.

Lean, John, 1818-77, railway engineer.
PICE - 52, 283-284.

Leather, John Towlerton, 1804-85, contractor(B).
PICE - 83, 433-436.
PIME - 1886, 262-263.

Leather, John Wignall, 1810-87, water works engineer(B).
Eng - 63, 107.
PICE - 89, 473-479.

Leather, Samuel Petty, 1821-89, water & gas engineer.
PICE - 98, 404.

Le Breton, Arthur Hemery, 1849-83, rly engineer overseas.
PICE - 75, 315.

Ledger, James Campbell, 1833-89, rly engineer in India.
PICE - 100, 388-389.

Lee, Henry, 1823-89, contractor.
PICE - 97, 422-424.

Leece, Joseph, 1833-86, armaments manufacturer(B).
PICE - 85, 399-401.

Lees, John, 1827-61, machinery manufacturer.
PIME - 1862, 17.

Lee-Smith, Hamilton, 1829-89, rly engineer overseas.
PICE - 99, 352-355.

Le Fanu, William Richard, 1816-94, Commissioner of Public
Works, Ireland(B).
PICE - 119, 395-397.

Legge, Cyrus, 18(13)-89, entrepreneur.
Elec - 23, 515*.

Leigh, Evan, 1810-76, inventor; consulting engr(B,D).
PICE - 44, 227-231.
PIME - 1877, 19-20.
TINA - 17, 345-347.

Leigh,Joseph Dorning,1822-78, engineering manufacturer.
PIME - 1879, 11.

Le Mesurier, Charles Benjamin, 1829-77, engineer in India.
PICE - 51, 266-268.

Le Mesurier, Henry Peveril, 1828-89, rly engr in India.
PICE - 98, 393-397.

Leonard, Moses,   -1880, iron works manager.
JISI - 17, 690-691.

Leslie,   ,   -1879, railway contractor.
Engng - 27, 165.

Leslie, Alexander, 1844-93, contractor.
Engng - 56, 730.
PICE - 116, 366-368.

Leslie, Andrew, 1818-94, shipbuilder.
PIME - 1894, 162-163.

Leslie, Frederick, 1828-89, gas engineer; inventor.
PICE - 98, 397-398.

Leslie, Hon. Henry Haworth, 1845-89, rly & canal engr overseas.
PICE - 98, 398-399.

Leslie, James, 1801-89, water engineer(B).
Engng - 49, 11.
PICE - 100, 389-395.

Lever, Charles, 1862-90, electrical engineer(B).
    Elec - 24, 245.
    Iron - 35, 33.

Levick, Frederick, 1803-67, ironmaster.
    PIME - 1868, 17.

Levinge, Harry Corbyn, 1828-96, civil engineer in India.
    PICE - 125, 404-405.

Lewis, Alfred David, 1837-99, shipbuilder.
    PICE - 139, 377-378.

Lewis, E   W   , 18(37)-1900, ironmaster.
    Eng - 89, 404*.

Lewis, Henry Watkin,   -1897, engineering manufacturer.
    JISI - 52, 255.

Lewis, John, 1825-65, mining engineer & surveyor.
    PICE - 25, 514-515.

Lewis, John Milton, 1846-81, railway engineer overseas.
    PICE - 70, 433-434.

Lewis, T   W   , 18(18)-1900, mechanical engineer.
    Eng - 89, 178.

Lewis, Thomas Lawrence, 1851-99, municipal surveyor.
    PICE - 139, 378-379.
    PIMCE - 25, 479-480*.

Leyland, Frederick Richards, 18(31)-92, telegraph company dir-
    ector & ship owner(B).
    Elec - 28, 250.

Liddell, Andrew, 1786-1855, ironmaster(B).
    PICE - 15, 102-103.

Liddell, Charles, 18(13)-94, railway contractor(B).
    Elec Rev - 35, 206.
    Eng - 78, 171.

Lindley, Robert Charles, 1824-87, civil engineer.
    PICE - 92, 404.

Lindley, William, 1808-1900, consulting municipal engr in
    Germany(B,D).
    Eng - 89, 541*.
    Engng - 69, 725.
    PICE - 142, 363-370.

Lindsell, Lt.-Col. John Barber, 1844-98, Royal Engineer.
    PICE - 133, 415.

Lindsley, Robert Riches, 1861-89, loco. engr in S.Africa.
TSE - 1889, 219.

Linging, Samuel Robert, 1851-85, gas engineer.
PICE - 83, 442.

Linton, F   T   C   ,   -1896, gas engineer.
Engng - 62, 804.

Lishman, William, 18(22)-98, mining engineer.
Eng - 85, 587.

Little, George, 1823-96, engineering manufacturer.
JISI - 50, 260.*
PIME - 1896, 600-601.

Little, Thomas David, 1842-1900, railway engr in India.
Eng - 89, 541*.
Engng - 69, 717.
PICE - 142, 370-371.

Livesey, Frank, 1844-99, gas engineer.
Eng - 87, 439.
Engng - 67, 623.
PICE - 137, 431-432.

Llewellin, David, 1826-80, civil & mining engineer.
PICE - 62, 356.

Llewellyn, William Hely, 1849-77, manufacturing engineer.
PIME - 1878, 12.

Lloyd, Edward John, 1827-94, canal engineer.
PICE - 120, 362-364.

Lloyd, John, 1825-89, mechanical engineer.
JISI - 34, 221.

Lloyd, Col. John Augustus, 1800-54, Surveyor-General of Maurit-
ius(B,D).
PICE - 14, 161-165.

Lloyd, John Horatio, 1788-1884, railway lawyer(B).
PICE - 78, 450-454.

Lloyd, Sampson, 1808-74, engineering manager(B).
Engng - 18, 274.
PICE - 39, 292-293.
PIME - 1875, 27-29.

Lloyd, Thomas, 1803-75, Chief Engineer to the Navy(B).
PICE - 41, 217-220.
TINA - 17, 342-343.

Loam, Matthew Hill, 1817-81, gas & water engineer.
PIME - 1882, 9.

Lobnitz, Henry C   , 1831-96, marine engineer.
Eng - 82, 661.
Engng - 62, 805.
TIESS - 40, 255.

Locke, Joseph, F.R.S., 1805-60, railway engineer(B,D).
Eng - 10, 194; 21, 61.
PICE - 20, 141-148.
TNEIMME _ 9, 56-62. By N. Wood.

Logan, David, 1832-96, railway engineer in India.
PICE - 125, 405-406.

Login, Thomas, 1823-73, hydraulic engineer in India(B).
Eng - 38, 52*.
PICE - 39, 269-271.

Londonderry, 5th Marquis of, 1821-84, industrialist(B).
Iron - 24, 451.

Longlands, Richard, 1830-85, railway engineer in India.
PICE - 88, 446-447.

Longridge, James Atkinson, 1817-96, contractor & arms manufr.
PICE - 127, 372-379.

Longridge, William Smith, 1819-78, ironmaster.
PIME - 1879, 11-12.

Longsdon, Alfred, 1827-93, Krupp's representative in U.K.
Engng - 56, 678, 698.
JISI - 45, 392*.
PICE - 116, 368-369.
PIME - 1893, 496.

Lonsdale, 3rd Earl of, 1818-76, industrialist(B).
JISI - 9, 474*.

Lord, Edward, 1812-75, inventor & manufacturer.
PIME - 1876, 22-23.

Lord, Thomas Wilkes, 1809-90, foundry owner.
PIME - 1871, 16.

Loutitt, Samuel Henry, 1838-93, water works engineer.
PICE - 116, 387-388.

Lovegrove, James, 1827-1900, municipal engineer.
PICE - 143, 334.

Lovell, Thomas, 1827-78, railway engineer in India.
PICE - 57, 309-311.

Low, George, 1833-94, design engineer.
    PIME - 1894, 276-277.

Low, Thomas Bell, 1855-86, civil engr in New Zealand(B).
    PICE - 91, 450-451.

Low, William,1814-86, colliery engineer(B).
    Eng - 62, 76.
    Iron - 28, 88.

Lowe, George, 1788-1868, gas engineer(B).
    PICE - 30, 442-445.

Lowe, John Edgar, 1839-97, engineering contractor.
    Engng - 84, 405.
    JISI - 52, 255-256.
    PICE - 131, 388.
    PIME - 1897, 235.

Lucas, Charles Thomas, 1820-95, contractor & builder(B).
    Engng - 60, 732.
    PICE - 124, 438-441.

Lumsden, James Ford, 18(69)-93,telegraph engr overseas.
    Elec - 31, 58*.
    Elec Rev - 32, 564.

Lundy, Thomas Evans,   -1872, electrical engineer.
    JSTE - 1, 139*.

Lunn, Robert Watson, 18(15)-90, hydraulic engineer.
    PICE - 100, 395-396.

Lunt, John, 1832-92, railway engineer in Australia.
    PICE - 112, 354-355.

Lüthy, Robert, 1840-84, draughtsman(B).
    Eng - 38, 68*.
    PIME - 1884, 403-404.

Lynam, Henry Middlemore, 1864-95, colliery manager.
    JISI - 49, 285-286*.

Lynde, James Gascoigne, 1816-83, municipal engineer.
    PICE - 73, 360-362.
    PIME - 1884, 66-67.
    PIMCE - 9, 206-207. Port. frontis v.18.

Lyons, John,   -1889, Public Works engineer in India.
    PICE - 101, 309-310.

Lysaght, St.John George,   -1892, manufacturer.
    Iron - 40, 472.

Lyster, George Fosbery, 1821-99, dock engineer, Liverpool(B).
Eng - 87, 497+port.
Engng - 67, 653.
PICE - 139, 357-366.

Macadam, Philip Henry, 1831-95, railway engineer overseas.
PICE - 123, 435-437.

McArthur, James, 18(38)-91, shipbuilder.
Engng - 51, 565.

McCallum, James Braddon, 1852-94, municipal engineer.
PICE - 119, 397-398.
PIMCE - 21, 326-328.

M'Clean, John Robinson, F.R.S., M.P., 1813-73, contracting
civil engineer(B).
Eng - 36, 38.
Engng - 16, 41.
PICE - 38, 287-291.

McClement, Donald, 18(67)-93, mechanical engineer.
TMAE - 3, 286.

McColl, Hugh, 1841-97, shipbuilder.
Eng - 83, 451.
NECIES - 13, 271-272.
TIESS - 41, 380.

McConnell, James Edward, 1815-83, consulting mechanical engr.
Eng - 55, 453.
PICE - 74, 285-286.

McConnochie, James Adair, 1835-95, dock engineer.
PICE - 125, 406-408.

McConnochie, John, 1823-89, dock engineer, Cardiff.
PICE - 97, 403.
PIME - 1889, 335-336.

McCowan, William, 18(38)-1900, ironmaster.
Eng - 89, 265*.
JISI - 57, 255.

McCrea, John Benjamin, 1833-95, contractor.
PICE - 122, 398.

M'Culloch, David, 1847-97, consulting mechanical engineer.
TIESS - 41, 380.
TSE - 1898, 233.

M'Cunn, John,   -1873, ship owner.
Engng - 16, 206*.

McCurrich, John Martin, 1852-99, dock engineer.
Eng - 87, 59.
Engng - 67, 79.
PICE - 136, 353-354.

M'Derment, J   I   ,   -1884, ---
TIESS - 28, 290*.

McDonald, Alexander, M.P., 1821-81, President, Miners'National
Union of England(B).
Eng - 52, 334.
Engng - 32, 467.
Iron - 18, 385.

McDonald, Alexander,   -1895, contractor.
Engng - 61,57.

Macdonald, John,   -1888, ____
TIESS - 31, 236*.

Macdonald, William,   -1899, civil engineer in Greece.
Eng - 87, 35*.

Macdougall, Alan, 1842-97, consulting engineer in Canada.
PICE - 129, 375-376.

McDowall, Daniel,   -1892, foundry owner.
Engng - 54, 666.

Macfarlaine, Thomas Milbourne, 1840-81, engineer in India.
PICE - 72, 316.

Macfarlane, James L   , 1858-93, marine engineer.
TIESS - 36, 322.

Macfarlane, Walter, 18(19)-85, ironmaster.
Engng - 40, 402, 431.
PIME - 1885, 526-527.
TIESS - 29, 220.

McGarel-Hogg, Sir James Macnaughten, Bart., Lord Magheramorne,
1823-90, Chairman, Metropolitan Board of Works.
PICE - 101, 311-312.

McGeogh, Andrew J   , 18(49)-99, founder and manufacturer.
Elec Rev - 44, 606.

Macglashan, John, 1842-84, railway engineer in India(B).
PICE - 80, 338-340.

Macglashan, William, 1865-95, railway engineer overseas.
PICE - 121, 334-335.

Macgregor, Donald Robert, 1824-89, ship owner and builder(B).
Engng - 48, 686*.

McGregor, Peter, 1860-96, marine engineer in China.
PIME - 1896, 601.

McGregor, William, 1819-91, engineering works foreman.
TMAE - 1, 200.

McIntosh, David, 1799-1856, contractor.
PICE - 16, 162-163.

McIntyre, John, 18(21)-1900, ship builder(B).
Eng - 89,239.

Maciver, Charles, 1812-85, ship owner.
Engng - 41, 42.
Iron - 27, 33.

Mackain, Daniel, 1800-59, water works engineer.
PICE - 19, 175-176.

Mackay, Edward, 1839-99, foundry manager.
TIESS - 43, 360-361.

Mackay, John, 1828-73, consulting engineer.
PIME - 1874, 22-23.

Mackenzie,   , 18(54)-91, electrical company secretary.
Elec - 27, 236*.

Mackenzie, Edward, 1811-80, civil engineer & contractor(B).
Engng - 30, 273.

Mackenzie, John, 1827-91, machinery manufacturer.
PICE - 106, 336-337.

Mackenzie, William, 1794-1851, contracting engineer.
CEAJ - 14, 584.
PICE - 11, 102-105.

McKerlie, Col. Sir John Graham, 1815-1900, Royal Engineer.
PICE - 140, 287-288.

Mackinlay, John, 1823-93, engineer in India.
PICE - 113, 350-351.

Mackinnon, Alexander Kendall, 1836-1900, civil engr in South
America.
PICE - 140, 276-277.

Mackinnon, Laughlan Alexander Entwistle, 1843-81, civil engr in
West Indies.
PICE - 65, 377.

Mackinnon, Sir William, Bart., 1823-93, ship owner(B,D).
Engng - 55, 899.

Mackintosh, Alexander, 1820-90, railway engineer.
PICE - 101, 299-300.

Mackison, Francis, 1822-84, civil engineer & architect.
PICE - 76, 360-361.

Mackworth, Hubert Francis, 1823-58, Inspector of Mines(B).
Eng - 6, 66*.
PICE - 18, 196-199.

Maclean, Sir Andrew, 1828-1900, ship builder(B).
Eng - 90, 517.
Engng - 70, 667.

McLean, John Robinson, M.P., 1813-73, consulting engineer(B).
PIME - 1874, 23-24.

Maclean, Loudon Francis, 1848-97, canal engr in India.
PICE - 133, 400, 402.

Maclellan, Walter, 1815-89, engineering manufacturer.
Engng - 47, 692; 48, 79*.
TIESS - 32, 322-333.

McLeod, Lt.-Gen. Duncan, 1780-1856, Bengal Engineer(B,D).
PICE - 16, 163-166.

McLeod, John, 18(24)-91, engineering works manager.
TMAE - 1, 199.

McMaster, Bryce, 1832-77, railway engineer in India.
PICE - 52, 277-278.

McMillan, Alexander, 18(51)-99, engineer in Japan.
Eng - 88, 634.

M'Millan, John, 18(15)-91, ship builder.
Eng - 72, 259.
Engng - 52, 363.
Iron - 31, 274.
TIESS - 35, 306-307.

M'Millan, John, Jun., 1848-88, shipbuilder & naval architect.
    Eng - 66, 477.
    Engng - 46, 551.
    Iron - 32, 503.
    JISI - 34, 218-219.

Macnab, Archibald F , -1899, engineer in Japan.
    Eng - 88, 634.

Macnab, James, 1872-1900, manufacturer.
    PIME - 1900, 627-628.

Macnab, Robert Aikenhead, 18(60)-1900, shipbuilder.
    Eng - 89, 265.

M'Nab, William, 18(20)-90, shipbuilder.
    Eng - 69, 468.

McNaught, William, 1813-81, steam engine manufacturer.
    Eng - 51, 53.
    Engng - 31, 66.

McNay, William, 1825-82, railway works manager.
    PIME - 1883, 19-20.

Macnee, Daniel, 1838-93, railway engineering contractor.
    JISI - 45, 393.
    PICE - 114, 381-382.
    PIME - 1893, 496-497.

Macneill, Sir John, F.R.S., 1793-1880, consulting civil engineer(B).
    Eng - 49, 197*, 215-216.
    Engng - 29, 203.
    Iron - 15, 190.
    PICE - 73, 361-371.

Macneill, Maj.-Gen. William Gibbs, 1801-53, railway engineer.
    PICE - 13, 140-145.

M'Onie, Andrew, 1818-86, manufacturing engineer.
    Engng - 41, 498*.
    TIESS - 29, 220.

McOnie, Sir William, 1813-94, machinery manufacturer(B).
    Eng - 77, 111.
    Engng - 57, 198.

McOnie, William, Jun., 1851-87, sugar machinery manufacturer.
    PIME - 1887, 277.
    TIESS - 30, 312.

McQueen, John, 1845-1900, engineering manufacturer.
    PIME - 1900, 333-334.

145

Macrae, John, 1836-93, consulting railway engineer.
PICE - 114, 382-384.

McRea, John Benjamin, 1833-95, railway contractor.
PICE - 122, 398.

Macritchie, James, 1847-95, municipal engineer, Singapore.
PICE - 122, 374-378.
PIMCE - 21, 329-330.
TIESS - 38, 335-336.

McVeagh, John,   -1861, civil engineer.
PICE - 21, 563-564.

Maddison, Ralph Rawling ,   -1878, mining engineer.
Eng - 46, 175.

Main, John, 18(47)-98, electrical engineer; inventor.
Elec Rev - 41, 340, 372.

Main, John Frederick, 1854-92; Prof. of Maths, Applied Mechanics
and Engineering, Bristol University College(B).
PICE - 110, 394-396.

Mainwaring, Henry, 1824-94, brass founder.
TMAE - 4, 228-229.

Majende, Col. Sir Vivian, 1836-98, Royal Engineer(B).
Eng - 85, 400.
Engng - 65, 528.

Makinson, Alexander Woodlands, 1822-86, civil engineer.
PICE - 86, 355.

Malkin, William Henry, 18(49)-94, brass founder.
TMAE - 4, 229*.

Mallaband, John, 18(30)-97, steel works manager.
Eng - 84, 239*.
Engng - 64, 260.
JISI - 52, 256.

Mallet, Robert, F.R.S., 1810-81, geologist & civil engr(B,D).
Eng - 52, 352-353, 371-372, 389-390.
PICE - 68, 297-304.

Mallinson, Thomas,   -1898, municipal surveyor.
PIMCE - 25, 479.

Manby, Charles, F.R.S., 1804-84, Secretary, Institution of
Civil Engineers(B,D).
Eng - 58, 110, 130.
Engng - 38, 158-159.
Iron - 24, 134-135.
PICE - 81, 327-334+ port. frontis.

Manby, Edward Oliver, 1816-64, ironmaster.
PICE - 24, 533-534.

Manby, John Richard, 1813-69, civil & mining engineer.
Eng - 27, 243.
PICE - 30, 446.

Manby, Joseph Lane, 1814-62, entrepreneur.
PICE - 22, 629-630.

Mann, Joseph Waddington, 1847-94, railway engineer.
PICE - 117, 392-393.

Manning, John, 1830-74, engineering manufacturer.
PIME - 1875, 29.

Manning, Robert, 1816-97, Public Works engr in Ireland.
PICE - 131, 370-371.

Manuel, David,   -1885, engineer in India.
PICE - 81, 339-340.

Mare, Charles John, 1814-98, shipbuilder(B).
Eng - 85, 133; 87, 83+port.

Margary, Peter John, 1820-96, railway engineer.
Engng - 61, 615.
PICE - 125, 409-410.

Margetson, A   G   , 18(50)-94, carriage works manager.
Engng - 58, 704.

Marillier, James Constantine, 1827-90, bridge engr overseas.
PICE - 102, 335-336.

Marindin, Col. Sir Francis Arthur, 1838-1900, Royal Engineer;
railway inspector.
Elec Rev - 46, 709-710.
Eng - 89, 437.
Engng - 69, 551.

Markham, Charles, 1823-88, coal & ironmaster.
Iron - 32, 218.
JISI - 33, 166-167.
PICE - 95, 377-379.
PIME - 1888, 439-440.

Marley, John, 1823-90, works manager.
Eng - 71, 287.
PICE - 105, 308-311.

Marsden, Benjamin, 1838-97, nut & bolt manufacturer.
Eng - 83, 153.
JISI - 51, 311-312.

PIME - 1897, 138.
TMAE - 7, 287-289.

Marsden, Henry Rowland, 1823-76, machinery manufacturer(B).
Eng - 41, 49.
Engng - 21, 71.

Marsden, John, 1849-86, colliery owner.
TCMCIE - 15, 32-33.

Marshall, Edwin, 1814-65, wagon manufacturer.
PICE - 25, 532*.
PIME - 1865, 15-16.

Marshall, Robert Alfred, 1852-84, colliery manager.
TCMCIE - 14, 65-66.

Marshall, William Ebenezer, 1824-80, foundry owner.
PIME - 1881, 5.

Marten, Henry John, 1827-92, hydraulic engineer.
JISI - 42, 296-297.
PICE - 112, 355-358.
PIME - 1892, 567-568.

Martin, Albinus, 1791-1871, surveyor & civil engineer(B).
PICE - 33, 223-226.

Martin, George,   -1887, colliery manager.
Engng - 43, 556*.

Martin, Henry Daniel, 1811-98, consulting civil engineer.
PICE - 135, 349-351.

Martin, Henry Robert Howells, 1852-87, gas engineer.
PICE - 91, 451-452.

Martin, Hon. James, 1821-99, engineering manufacturer.
Eng - 89, 181.
PIME - 1900, 332-333.

Martin, Samuel Dickinson, 1803-77, inventor & surveyor(B).
Eng - 41, 216*.
PICE - 53, 292-293.

Martin, Thomas, 1831-83, Public Works engineer in India.
PICE - 78, 435-436.

Martin, Thomas George, 1849-96, naval architect.
PIME - 1896, 258-259.

Martin, Walter Richards, 1860-93, civil engineer.
PICE - 116, 383.

Martineau, Louis, 1866-95, civil engineer overseas.
PICE - 122, 399.

Martley, William, 1824-74, locomotive superintendent.
Eng - 37, 126.
PICE - 41, 221.
PIME - 1875, 29-30.

Massey, Frederick Henry, 1812-97, civil engineer overseas(B).
Eng - 84, 573.

Masterson, Henry, 1843-98, civil engineer.
PICE - 135, 367.

Mather, William Ernest, 18(77)-99, ---
Eng - 88, 475, 483.

Mathew, Francis, 1847-85, railway engineer.
PICE - 83, 436-437.

Mathews, William, 1796-1871, ironmaster.
PIME - 1872, 19-20.

Mathias, James, 1823-86, railway engineer.
PICE - 86, 356-357.

Matthew, John, 1819-69, draughtsman.
PICE - 30, 446-448.

Matthews, William Anthony, 1813-72, steel manufacturer(B).
PIME - 1873, 18-19.

Maudslay, Henry, 1822-99, engineering manufacturer.
Eng - 87, 81+port.
Engng - 68, 110.
PICE - 138, 490-491.
PIME - 1899, 270.

Maudslay, Joseph, 1801-61, marine engr & shipbuilder(B,D).
Eng - 12, 216*.
Mech Mag - 56, 250, 351-352.
PICE - 21, 560-563.
TINA - 2, 19.

Maughan, James Archibald, 1843-94, mining engineer in India.
PICE - 121, 335-336.

Maunsell, Frederick William, 1859-94, Public Works engr in India.
PICE - 118, 461-462.

Mavor, Ivan, 18(59)-86, shipbuilding works manager.
NECIES - 3, xxxiii-xxxiv.
TIESS - 30, 314.

Mavor, Percy William, 1851-99, mining engineer.
PICE - 139, 379.

Maxwell, William Henry, 1803-87, draughtsman.
PICE - 91, 467.

Maxwell, William John Leigh, 1838-80, railway engineer
overseas(B).
PICE - 66, 384-385.

May, Charles, F.R.S., 1800-60, engineering manufacturer(B).
Eng - 10, 112*.
PICE - 20, 148-149.

May, George, 1805-67, Caledonian Canal Engineer.
PICE - 27, 595-596.

May, James, 1818-75, Caledonian Canal Engineer.
PICE - 45, 244-245.

May, Robert Charles, 1829-82, consulting engineer(B).
Eng - 54, 89.
Engng - 34, 118.
Iron - 20, 87.
PICE - 73, 367-368.
PIME - 1883, 20.

May, Walter, 1831-77, engine manufr & iron founder.
PIME - 1878, 12-13.

Mayes, William Frederick, 1848-84, railway engr overseas.
PICE - 82, 386-387.

Maylor, John, 1827-87, mechanical engineer in Brazil.
JISI - 28, 219.
PIME - 1887, 539.

Mead, Lt.-Col. Clement John, 1832-76, Bengal Engineer.
PICE - 47, 302-304.

Mead, Frank, 1852-97, gas & water works engineer.
PICE - 131, 388-389.
TSE - 1897, 202.

Meara, Thomas Francis, 1837-97, contractor overseas.
PICE - 132, 385-386.

Medley, Maj.-Gen. Julian George, 1829-84, Royal Engineer(B).
PICE - 80, 343-347.

Meek, Sturges, 1816-88, chief railway engineer.
Eng - 65, 181.
PIME - 1888, 155-156.

Meik, Thomas, 1812-96, consulting dock engineer.
  PICE - 125, 410-412.
  PIME - 1896, 96-97.

Meixner, Frederick,   -1895, ---
  TMAE - 5, 265*.

Melson, Dr John Barrett, 1811-98, electrician(B).
  Elec Rev - 42, 774.
  Eng - 85, 529.

Menelaus, William, 1818-82, ironworks manager(B).
  Engng - 33, 347, 398.
  Iron - 19, 250.
  JISI - 21, 633-635.
  PIME - 1883, 20-22.

Menzies, William, 1840-98, consulting engr & marine surveyor.
  NECIES - 15, 265-266.
  PIME - 1898, 537-538.
  TIESS - 42, 409.

Meredith, George, 1817-64, railway engineer.
  PICE - 25, 515-516.

Merrifield, Charles Watkins, F.R.S., 1828-83, marine engr &
  naval architect(B).
  Eng - 57, 53.
  TINA - 25, 309-310.

Merry, James, 1805-77, ironmaster & colliery owner(B).
  Engng - 23, 117.
  JISI - 9, 512*.

Merryweather, Henry, 1846-81, fire engine manufacturer.
  Eng - 53, 7.
  Engng - 33, 18.
  PIME - 1882, 9.

Merryweather, Moses, 1791-1872, fire engine manufacturer(B).
  Engng - 14, 234.

Merryweather, Richard Moses, 1839-77, fire engine manufr(B).
  PIME - 1878, 13.

Messent, Philip John, 1830-97, river engineer.
  Eng - 83, 371.
  Engng - 63, 477*, 519-520.
  PICE - 129, 376-379.

Metford, William Ellis, 1824-99, arms inventor(B).
  Eng - 88, 392, 432.
  PICE - 140, 288-293.

Methven, John, 1845-98, gas works engineer.
Engng - 65, 472.
PICE - 133, 402-403.

Méyer, Christian Hendrick, 1830-87, civil engr & surveyor.
PICE - 92, 399-400.

Michell, William, 1853-99, railway engineer in India.
PICE - 139, 379-381.

Michell, William Marwick,   -1885, patent office official.
Eng - 59, 238.

Middlemas, William,   -1894, iron ore merchant.
Engng - 58, 834-835.

Middleton, Thomas, 1819-91, railway & dock engineer.
Eng - 71, 169.

Middleton, William, 1795-1868, foundry owner.
PIME - 1869, 16.

Midelton, Thomas, 1848-94, patent expert.
PIME - 1894, 277-278.

Miers, John William, 1819-92, civil engineer in Brazil.
Engng - 53, 199.
PICE - 109, 429-430.
PIME - 1892, 99.

Miles, Thomas William, 1840-95, Public Works engr in India.
PICE - 122, 378-379.

Millar, William, 1814-87, engineer in Russia.
PICE - 91, 452.

Miller, Alexander, 18(27)-91, ironmaster in India.
Iron - 38, 454.

Miller, Daniel, 1826-88, consulting dock engineer.
Engng - 46, 363.
PICE - 96, 322-333.
TIESS - 32, 323.

Miller, George, 1802-84, railway contractor.
Iron - 23, 337.

Miller, George Mackay, 1813-64, chief railway engineer.
Eng - 17, 183*.
PICE - 24, 534-536.
PIME - 1865, 16.

Miller, John, 1805-83, railway contractor(B).
Eng - 55, 366.

```
        Engng - 35, 493-494.
        PICE - 74, 286-289.

Miller, Joseph, F.R.S., 1797-1860, marine engr(B).
        PICE - 20, 149-156.

Milligan, Robert, 1827-76, rly engineer in South America.
        PICE - 47, 298-299.

Mills, George, 1796-1859, mechanical & marine engineer.
        PICE - 19, 191-192.

Mills, William, 1818-91, chief railway engineer(B).
        PICE - 108, 405-406.

Milroy, John, 1806-86, railway pioneer.
        Engng - 42, 431.
        PICE - 88, 454-456.

Milward, Col. Thomas Walter, 1826-74, Royal Artillery(B).
        Eng - 39, 27.

Minto, George, 1825-79, mining engineer.
        Eng - 48, 79.
        Engng - 28, 99.
        TCMCIE - 8, 13-17.

Mitchell, Alexander, 1807-48, civil engr & surveyor.
        PICE - 8, 103.

Mitchell, Charles, 1820-95, shipbuilder.
        Eng - 80, 213.
        Engng - 60, 276.
        NECIES - 12, 248-249.

Mitchell, Edwin, 18(46)-94, foreman moulder.
        TMAE - 4, 229*.

Mitchell, Joseph, 1803-83, road engr in the Highlands(B).
        Engng - 36, 523-524.
        Iron - 22, 511.
        PICE - 76, 362-368.

Mitchell, Joseph, 18(07)-76, colliery owner.
        Eng - 41, 49.

Mitchell, Joseph, Jun., 1840-95, mining engr & coal owner.
        Eng - 79, 359.
        Engng - 59, 542.
        JISI - 47, 262-263.
        PICE - 121, 325-327.
        TIME - 10, 107-108. By G.J. Kell.
```

Mitchell, Thomas Telford, 1815-63, rly engr & contractor.
PICE - 24, 542-543.

Mitchell, William,   -1849, marine engineer.
PICE - 9, 104*.

Moffat, Thomas, 1830-86, mine manager.
PIME - 1887, 147-148.

Moir, James, 1855-1900, superintending marine engineer.
PIME - 1900, 628.

Mole, Thomas,   -1856, sword manufacturer.
Eng - 1, 218*.

Molesworth, Robert Bridges, 1863-97, civil engineer.
PICE - 131, 371-372.

Monckton, Claud, 1844-97, civil engineer.
PICE - 131, 389-390.

Montgomerie, Lt.-Col. Patrick, 1837-86, Royal Engineer(B).
PICE - 86, 368.

Montgomery, John James, 1832-84, municipal engineer(B).
PICE - 78, 436-439.

Montrésor, Charles Edward Cage, 1857-91, Public Works engr in
India.
PICE - 106, 345-346.

Moodie, William, 1811-72, inventor of screw propeller.
Engng - 13, 160*.

Moody, Maj.-Gen. Richard Clement, 1813-87, Royal Engineer(B).
PICE - 90, 453-455.

Moon, Sir Richard, Bart., 1814-99, railway contractor(B).
Eng - 88, 524, 577.
Engng - 68, 658.

Moore, Alexander, 1809-78, naval dockyard instructor(B).
Eng - 45, 246.

Moore, Benjamin Theophilus, 1830-99, Prof.of Civil Engng & Applied
Mechanics, University College, London(B).
PICE - 139, 366-367.
PIME - 1899, 618.

Moore, James Michael, 1856-91, civil engineer.
PICE - 107, 418-419.

Moore, Ralph, 1822-96, Inspector of Mines.
TIESS - 40, 255-257.

Moore, Thomas, 18(36)-91, telegraph company director.
    Elec - 26, 414.
    Elec Rev - 43, 755*.

Moore, William, 1834-89, mining engineer.
    PICE - 97, 404.

Moorsom, Capt. William Scarth, 1804-63, railway engineer(B).
    Eng - 15, 335*.
    PICE - 23, 498-504.

Morant, Alfred William, 1828-81, municipal engineer(B).
    Iron - 18, 128.
    PICE - 66, 377-379.
    PIMCE - 7, 145-146+ port.frontis. v.19.

Morant, Ernest Frederick, 1849-88, rly engr in S.America.
    PICE - 95, 387-388.

Morant, Major James Law Lushington, 1839-86, Royal Engr.
    PICE - 80, 370-373.

Morant, John Hemphill, 1847-92, railway engr in Brazil(B).
    PICE - 110, 380-382.

Mordue, G   R   ,   -1874, ironworks manager.
    Eng - 37, 171*.

Morel, Edmund, 1841-71, consulting engineer overseas.
    PICE - 36, 299-300.

Moreland, Richard, 1805-91, engineering manufacturer.
    PICE - 104, 316-318.

Morfee, Capt. C   Bulmer, 18(58)-93, telegraph engineer.
    Elec - 31, 4.

Morgan, Charles,   -1892, brass & iron founder.
    Iron - 39, 570*.

Morgan, Charles William, 1859-86, civil engineer.
    PICE - 87, 448-449.

Morgan, David,   -1900, miners' agent.
    Eng - 90, 39.

Morgan, James, 1837-99, municipal engineer.
    PICE - 139, 381.

Morgan, John, 18(08)-(93),   ---
    TIESS - 37, 197.

Morgan, John, 1835-1900, railway company secretary(B).
    Eng - 90, 714*.

Morgan, Joseph Bond, 1834-1900, entrepreneur.
    Elec Rev - 46, 952.
    PICE - 142, 394-395.

Morgan, Joshua Llewellyn, 1820-75, ironmaster.
    PIME - 1876, 23-24.

Morgan, Octavius Vaughan, 1837-96, manufacturer(B).
    JISI - 49, 287.

Morgan, Thomas Rees, 1834-97, engineering manufr in USA.
    JISI - 52, 256.
    PIME - 1898, 315-316.

Morgan, William, 18(34)-97, electrical engineer.
    Elec Rev - 41, 312.

Moriarty, Edward Orpen, 1824-96, consulting civil engr in NSWales
    PICE - 129, 379-381.

Morley, Richard, 1823-91, ironworks manager.
    TMAE - 1, 200-201.

Morris, Claude John, 18(32)-1900, wire manufacturer.
    JISI - 58, 391.

Morris, W   R   ,   -1891, telephone inspector.
    Elec Rev - 29, 140*.

Morris, William, 1836-86, civil engineer.
    PICE - 87, 422-423.

Morris, William Richard, 1808-74, water works engineer(B).
    Eng - 37, 42*.
    PICE - 39, 271-273.

Morrison, James, 1806-78, ironmaster.
    Engng - 25, 122*.
    JISI - 12, 290-291.

Morrison, Martin,   -1900, coal & ironmaster.
    Eng - 89, 157-158.
    Engng - 69, 187.

Morrison, Robert, 1822-69, manufacturer(B).
    PICE - 31, 220-222.

Morrow, William Henry, 1844-94, surveyor.
    PICE - 119, 407-408.

Morton, Andrew, 1847-99, railway engineer in India.
    PICE - 142, 371-372.

Morton, Charles, 18(11)-82, Inspector of Mines.
    Iron - 20, 421.

Morton, Henry Thomas, 1825-98, agent for Earl of Derby.
    Eng - 86, 22.

Morton, James,   -1896, company secretary.
    Elec Rev - 43, 462.

Morton, John,   -1892, railway company manager.
    Eng - 74, 55.

Morton, William, 1832-99, railway contracting engineer.
    PICE - 136, 355.

Moseley, Charles, 1840-87, elec. engr & rubber manufr(B).
    Elec - 19, 455.
    Elec Rev - 21, 370.
    Eng - 64, 296.
    Iron - 30, 353-354.

Moseley, Canon Henry, F.R.S., 1801-72, naval architect(B).
    TINA - 13, 328-330.

Mostyn, Piers,   -1885, factory inspector.
    Engng - 39, 188*.

Moylan, William,   -1892, mech. engr; Inspector of Boilers.
    Iron - 39, 163.

Mudd, Thomas, 1852-98, marine engineer.
    Eng - 85, 483, 491.
    Engng - 65, 639, 669.
    JISI - 53, 318.
    NECIES - 15, 266-267.
    PICE - 134, 405-406.
    PIME - 1898, 538-539.

Muir, James, 1817-89, consulting water engineer(B).
    PICE - 96, 323-326.

Muir, Matthew Andrew, 1812-80, foundry owner(B).
    PICE - 61, 298.

Muir, William, 1806-86, marine engineer.
    Eng - 66, 167-168.
    Engng - 42, 351-352.

Muir, William, 1806-88, machine tool manufacturer(B).
    Engng - 46, 194.
    PIME - 1888, 440-442.

Muirhead, James, 1807-98, iron founder.
    Eng - 86, 451.
    Engng - 66, 594.

Muirhead, John, 1807-85, electrical engineer.

Eng - 60, 268.
Engng - 40, 322.
Iron - 26, 310.

Muirhead, John, 1846-95, electrical engineer.
Elec - 36, 177*.
Elec Rev - 37, 680*.
PICE - 126, 399-400.

Mulock, George Phillips, 1851-98, railway engineer.
PICE - 133, 403-404.

Munday, George James, 1820-97, contractor.
PICE - 132, 399-400.

Mungall, James, 18(12)-97, colliery owner.
JISI - 53, 318-319.

Murdoch, Sir George,   -1889, Chief Inspector of Machinery
to the Navy.
Eng - 67, 30.

Murdoch, James, 1821-98, shipbuilder & forge owner.
TIESS - 42, 410.

Murdock, William, 1754-1839, gas engineer.
CEAJ - 13, 386-387.

Muriel, George Brooke, 1842-81, gas engineer.
PICE - 65, 377-378.

Murphy, Henry de la Poire, 1828-71, railway engr in India.
PICE - 33, 265-267.

Murray, Andrew, 1813-72, consulting engr to the Navy(B).
Eng - 34, 264*.
Engng - 14, 271.
PICE - 36, 270-273.
TINA - 14, 232-234.

Murray, Edward Francis, 1818-82, railway engineer.
PICE - 74, 289-290.

Murray, John, 1804-82, hydraulic engineer(B).
PICE - 71, 400-407.

Murray, Maj.-Gen. Robert, 18(29)-89, Director-General of Telegr-
aphs in India.
Elec - 23, 595.

Murray, Thomas,   -1860, mechanical engineer.
Eng - 10, 309*.

Murray, Thomas Hunter, 1824-78, machinery manufr.
    PIME - 1879, 12.

Murton, Frederick, 1817-89, rly engineer & contractor(B).
    PICE - 96, 326-328.

Musgrave, James, 18(26)-96, engineering manufacturer.
    Engng - 62, 318.

Musgrave, John, 1820-89, machinery manufacturer.
    Eng - 67, 305.
    JISI - 34, 218.

Musgrave, Joseph, 1812-91, ironfounder.
    Eng - 71, 53.
    Engng - 51, 73.
    JISI - 39, 239-241.
    PICE - 104, 313-315.

Mushet, David,   -1847, ironmaster.
    PICE - 7, 11-12.

Mushet, Robert Forester, 1811-91, metallurgist(B).
    Elec - 26, 446.
    Eng - 71, 106.
    Engng - 51, 189.

Mylne, William Chadwell, F.R.S., 1781-1863, surveyor(D).
    PICE - 30, 448-451.

Mylne, William Chadwell,Jun., 1821-76, water works engr(B).
    PICE - 44, 231-232.

Nabholz, Karl Emil, 1856-94, draughtsman.
    PICE - 120, 372-373.

Napier, Charles George, 1829-82, consulting engineer.
    Eng - 54, 190.
    PICE - 71, 407-408.

Napier, David, 1790-1869, shipbuilder & marine engineer(B).
    Eng - 29, 102.

Napier, Francis, 1820-75, marine engineer.
    PICE - 44, 232-233.

Napier, George, 1869-98, admiralty dock engineer.
    PICE - 135, 368.

Napier, James, 1810-84, industrial chemist(B).
    Engng - 38, 553-554.

Napier, James Murdoch, 1823-95, hydraulic equipment manufr.
    PICE - 122, 379-382.
    PIME - 1895, 539-540.

Napier, James Robert, F.R.S., 1821-79, shipbuilder & marine
    engineer(B).
    Engng - 28, 497.
    TINA - 21, 265-266.

Napier, John, 1823-1900, engineering manufacturer.
    Engng - 69, 16.
    PICE - 140, 278-279.
    TIESS - 43, 361-363.

Napier, John, 1832-83, engineer in India.
    PICE - 80, 340-341.

Napier, John D   , 18(20)-80, marine engine manufacturer.
    Engng - 29, 231.

Napier, Richard John, 18(59)-91, Lloyd's surveyor.
    Engng - 51, 219.
    NECIES - 7, xlix-1.

Napier, Robert, 1791-1876, marine engr & shipbuilder(B).
    Eng - 41, 491.
    Engng - 21, 554-555. Port. v.4, 597.
    PICE - 45, 246-251.
    PIME - 1877, 20-21.
    Prac.Mag. - 3, 1,6-8.

Napier, Robert D   , 1826-85, mechanical engr & shipbuilder(B).
    Eng - 59, 387.
    Engng - 39, 584.
    Iron - 25, 453.

Nasmyth, James, 1808-90, inventor of steam hammer.
    Elec - 25, 2*.
    Eng - 69, 383, 406, 426.
    Engng - 49, 569.
    Iron - 35, 404.

Nasmyth, John, 18(22)-92, mechanical engineer.
    TMAE - 2, 279.

Naylor, John William, 1827-99, engineering manufacturer.
    JISI - 56, 266-267.

Neale, Deodatus Hilin, 1849-93, locomotive engineer.
    Eng - 75, 342.
    Iron - 41, 339.
    PICE - 113, 354-355.

Needham, Edward Moore, 1819-90, railway superintendent.
    Eng - 69, 99.

Needham, John,   -1899, engineering manufacturer.
    JISI - 56, 294-295.

Needham, Joseph, 18(33)-98, manufacturer.
    Engng - 65, 663*.

Neilson, Hugh, 1820-84, iron manufacturer.
    Engng - 39, 12*.

Neilson, Hugh,   -1890, iron & steel manufacturer.
    Engng - 49, 676*.
    TIESS - 34, 324.

Neilson, James Beaumont, F.R.S., 1792-1865, ironmaster(B,D).
    Eng - 19, 70.
    PICE - 30, 451-453.

Neilson, Walter, 1807-84, ironmaster(B).
    Engng - 38, 198.
    JISI - 27, 533-536.
    PICE - 80, 347-349.
    TIESS - 28, 290.

Neilson, Walter, Jun.,   -1896, ironmaster.
    JISI - 49, 287-288*.
    TIESS - 39, 271.

Neilson, Walter Montgomery, 1819-89, locomotive builder(B).
    Eng - 68, 33.
    Engng - 48, 54, 79.
    Iron - 34, 54.
    PICE - 100, 400-401.

Neilson, William, 1810-82, ironmaster & manufacturer.
    Engng - 33, 563.
    JISI - 21, 655-657.

Nelson, Thomas, 1807-90, builder & railway contractor(B).
    Engng - 50, 404.

Nelson, Thomas Boustead, 1842-78, railway engineer.
    PICE - 54, 284-285.

Ness, Walter, 1834-86, mining engineer.
    Iron - 27, 51.

Nethersole, William, 1829-95, railway engineer in India.
    PICE - 124, 416-417.

Nettell, E   C   ,   -1899, iron works manager.
    Eng - 87, 478*.

Nettlefold, Hugh, 1858-93, engineering manufacturer.
    PICE - 116, 383-384.

Nettlefold, Joseph Henry, 1827-82, inventor & manufacturer(B).
    Iron - 18, 443.
    PIME - 1882, 9-10.

Neumann, George, 1817-98, railway engr in Europe.
    Eng - 86, 133*.
    PICE - 134, 406-407.

Neville, Henry James Walton, 1822-74, admiralty engineer.
    PICE - 40, 259-260.

Neville, Park, 1812-86, municipal engineer(B).
    PICE - 87, 424-427.

Neville, Richard,   -1892, iron & tin manufacturer.
    Iron - 39, 516-517.

Newall, Robert Stirling, F.R.S., 1812-89, cable manufr(B,D).
    Elec - 22, 728.
    Elec Rev - 24, 510.
    Eng - 67, 353.
    Engng - 47, 410.
    Iron - 33, 363.
    PIME - 1889, 336-337.
    TIESS - 32, 324.

Newbold, Robert,   -1896, manufacturer.
    Eng - 82, 47.

Newcombe, Edward, 1843-86, railway engineer.
    PICE - 86, 357-358.

Newdigate, Albert Lewis, 1840-88, harbour engineer.
    PICE - 92, 400-401.
    PIME - 1888, 263-264.

Newdigate, Edward, 1866-96, municipal engineer.
    PICE - 127, 390-391.

Newham, William Edward, 1857-97, Public Works engr in India.
    PICE - 132, 396-397.

Newlands, James, 1813-71, City Engineer, Liverpool(B).
    Eng - 32, 40.
    PICE - 33, 227-231.

162

Newman, Edward, R.N., 1832-82, admiralty dock engineer(B).
Eng - 54, 426.
Engng - 34, 548.
Iron - 20, 486.

Newman, Frederick, 1837-86, water engineer.
PICE - 88, 447-448.

Newton, Alfred Howard Vincent, 1852-97, municipal engr overseas.
PICE - 130, 315-316.

Newton, Alfred Vincent, 18(19)-1900, patent agent.
Eng - 89, 647*.

Newton, John, 1829-96, consulting municipal engineer.
PICE - 127, 381-382.
PIMCE - 23, 479.

Newton, William, 1786-1861, patent agent & draughtsman(B).
PICE - 21, 592-594.

Newton, William Edward, 1818-79, patent agent.
PIME - 1880, 8.

Nichol, Peter Dale, 1831-71, engineering manufacturer.
PIME - 1872, 20.

Nicholas, Evan, 1842-95, mechanical engineer.
JISI - 48, 352*.

Nicholls, Theophilus, 1835-83, railway engineer.
PICE - 76, 368-369.

Nichols, John,   -1893, brass & iron founder.
Iron - 41, 88*.

Nicholson, John, 1812-66, marine superintendent.
Eng - 21, 421.*.

Nicholson, Robert, 1808-55, railway engineer.
PICE - 15, 93-94.

Nicholson, William Newman, 1816-99, agricultural machinery
manufacturer(B).
Eng - 87, 491.

Nicol, Bryce Gray,   -1892, engineering manufacturer.
NECIES - 8, 305-306.

Nicoll, Donald, 1820-91, inventor(B).
PICE - 108, 411-412.

Nisbet, William David, 1837-97, harbour engr in Queensland.
PICE - 121, 373-374.

Nixon, Charles, 1814-73, railway engineer.
   PICE - 38, 291-293.

Nixon, John, 1815-99, coal owner(B).
   Eng - 87, 595.
   Engng - 67, 750.

Norris, Moraston Ormerod, 1857-85, Public Works engr in India.
   PIME - 1886, 121-122.

Norris, Richard Stuart, 1812-78, railway official.
   PICE - 54, 281-283.
   PIME - 1879, 12-13.

North, John Thomas, 1842-96, mechanical engr & entrepreneur(B).
   Eng - 81, 471.
   JISI - 49, 288-289.
   PIME - 1896, 259.

Norton, William Baron, 1862-98, municipal engineer.
   PICE - 137, 442.

Norwood, Charles Morgan, M.P., 1825-91, ship owner(B).
   Iron - 37, 386.

Noyes, Philip Algernon Herbert, 1843-75, mechanical engineer.
   PICE - 44, 233-234.

Nuttall, Thomas, 1838-1900, consulting civil engineer.
   Eng - 89, 487*, 497.
   PICE - 141, 352-353.
   PIMCE - 26, 252.

Nye, John Henry, 1830-76, manufacturer.
   PIME - 1877, 21.

Oakes, Thomas, 18(08)-91, ironmaster.
   Iron - 37, 473.

O'Brien, W F ,   -1898, telegraph engr overseas.
   Elec Rev - 42, 19.

Ogilvie, Alexander, 1812-86, railway contractor.
   PICE - 86, 373-374.

Ogilvie, Arthur Graeme, 1851-97, colliery owner.
Eng - 84, 144*.
JISI - 52, 256-271.
PICE - 130, 319.

Ogilvie, Charles Edward Walker, 1823-90, railway engineer.
PICE - 103, 376-371.

Ogilvie, F   Douglas Walker, 18(71)-98, ---
Eng - 86, 375*.

Ogilvie, Robert, 1815-75, railway engineer.
PICE - 42, 263-264.

Ogle, Richard, 18(24)-99, iron merchant.
Eng -87, 21.
JISI - 56, 267.

O'Hagan, Henry, 1819-69, surveyor & railway engineer.
PICE - 30, 453-454.

Okell, James Smith, 1834-77, railway engr in South America.
PICE - 49, 276.

Olander, Edmund, 1834-1900, railway engineer.
Eng - 90, 436.
Engng - 70, 567.
PICE - 143, 318-320.

Oldham, James, 1801-90, civil engineer(B).
Eng - 69, 504.
PICE - 103, 377-380.

Oldham, John,   -1840, engineer to the Bank of England.
CEAJ - 4, 127.
PICE - 1, 14-15.

Oldham, Samuel,   -1892, consulting engineer.
Iron - 40, 231.

Oliphant, James, 18(12)-1900, engineering manufacturer.
Engng - 69, 16.

Oliver, Elihu Henry, 1839-76, municipal engr in Shanghai.
PICE - 45, 256-257.

Oliver, Robert Stewart, 1849-1900, surveyor.
PICE - 142, 389.

Olrick, Harry, 18(51)-86, construction engineer.
Engng - 42, 259*.
Iron - 28, 239.

Olrick, Lewis, 1827-80, mechanical engineer.

Eng - 50, 155.
JISI - 17, 691.
PIME - 1881, 6.

Olver, Joseph Salter, 1818-86, civil engineer.
PICE - 86, 366-367.

O'Meara, Patrick, 1834-98, railway engineer overseas.
PICE - 132, 383-385.

O'Meara, Thomas Francis, 1837-97, railway engr overseas.
PICE - 132, 385-386.

Ommanney, George Willes, 1848-87, locomotive engr in Trinidad.
PICE - 90, 443.

Ord, J   R   ,   -1900, agricultural implement manufacturer.
Eng - 90, 190.

Orde-Browne, Capt. Charles, 1838-1900, ordnance engineer.
Eng - 90, 263+port.
Engng - 70, 310.

Ordish, Rowland Mason, 1824-86, contracting engineer(B).
Eng - 62, 232-233; 63, 220.
Engng - 42, 298.
Iron -28, 265.

Orlebar, Cuthbert Knightley, 1844-82, railway engineer.
PICE - 71, 408-409.

Ormiston, Thomas, 1826-82, dock engineer(B).
PICE - 71, 409-414.
PIME - 1883, 22-24.

Ormsby, Arthur Sydney, 1825-87, railway & water engineer(B).
PICE - 89, 479-481.

Orwin, Charles Herbert, 1853-91, telegraph engineer.
Elec - 27, 653*.
Elec Rev - 29, 427.

Osborn, Samuel, 1826-91, steel manufacturer(B).
Iron - 31, 32.
PIME - 1891, 291-292.

Osborn, Admiral Sherard, 1822-75, hydrographer(B).
Eng - 39, 342.

Osborne, Richard Boyse, 1815-1900, civil engineer(B).
Eng - 89, 34.
Engng - 69, 13.

Osborne, Thomas, -1892, manufacturer.
    Iron - 39, 517.

Oswald, Ralph Patterson William, -1897, Inspector of Mines.
    Eng - 83, 605.

Otter, Francis William, 1847-85, railway engr overseas.
    PICE - 80, 341-342.

Ottley, Drewry Gifford, 1845-96, Public Works engr in India.
    PICE - 127, 382-383.

Owen, B , -1873, railway secretary.
    Eng - 36, 82*.

Owen, Charles Thompson, 1843-82, draughtsman & inventor.
    TCMCIE - 11, 13-14.

Owen, Dyson, 18(43)-95, coal & ironmaster.
    Eng - 79, 187, 201.

Owen, G Wells, 18(40)-95, railway engineer.
    Engng - 59, 286.

Owen, Thomas Ellis, 1841-1900, railway engineer in India.
    PICE - 142, 389-390.

Owen, William, 1810-81, foundry owner(B).
    PICE - 63, 333.
    PIME - 1882, 10-11.

Ower, Charles, 18(16)- 76, consulting harbour engineer.
    Engng - 22, 281.

Oxley, James Abbott, 1852-95, railway engineer overseas.
    PICE - 121, 336-337.

Oxley, William Henry, -1892, steel manufacturer.
    Iron - 39, 476.

Paddison, George, 1825-63, railway engineer.
    PICE - 31, 222-224.

Paddon, William Vye, 1859-90, military engr in Egypt.
    PICE - 103, 386-387.

Pagan, John, 1842-88, surveyor in Gold Coast(B).
    PICE - 96, 348-349.

Page,Christopher,1853-88, railway engineer.
    Engng - 45, 263*.

Page, George Gordon, 1836-85, surveyor.
    PICE - 82, 377-378.

Page, George Thomas, 1808-49, dock engineer.
    PICE - 10, 91.

Page, John,   -1885, consulting engineer.
    TIESS - 29, 220-221.

Page, Thomas, 1803-77, civil engineer(B).
    Eng - 42, 31.
    Engng - 23, 75.
    PICE - 49, 262-265.

Paget, Arthur, 1832-95, textile machinery manufacturer.
    Engng - 59, 436-437.
    PIME - 1895, 144-145.

Palliser, Sir William, M.P., 1830-82, military inventor(B).
    Eng - 53, 105-106.
    Engng - 33, 141.
    Iron - 19, 106.
    PICE - 69, 418-421.

Palmer, Henry Robinson, F.R.S., 1795-1844, civil engineer.
    CEAJ - 7, 371*; 8, 126.
    PICE - 4, 6-8.

Palmer, Maj.-Gen. Henry Spencer, 1838-93, Royal Engineer(B).
    PICE - 113, 373-375.

Park,   Edward Matthew, 1850-84, engineer in Brazil.
    PICE - 77, 383-384.

Park, James Crawford, 1838-95, locomotive superintendent.
    PICE - 122, 382.

Park,   John Carter, 1822-96, mechanical engineer overseas.
    PICE - 127, 383-384.

Parker, Maj. Francis George Shirecliffe, 1836-90, Public Works
    engineer in India.
    PICE - 100, 415-416.

Parker, W R ,   -1898, contractor.
    Eng - 85, 24*.
    Engng - 65, 24.

Parkes, Alexander, 1813-90, inventor(B).
    Elec - 25, 274.
    Eng - 70, 65-66.
    Engng - 50, 111.

Parkes, Henry Persehouse, 18(32)-1900, manufacturer.
    JISI - 57, 255.

Parkes, Josiah, 1793-1871, agricultural engineer(B).
    Eng - 32, 122.
    PICE - 33, 231-236.

Parkes, William, 1822-89, civil & docks engr overseas(B).
    PICE - 96, 328-330.

Parkin, Joseph,   -1869, iron & steel merchant.
    Eng - 28, 39*.

Parry, Albert Woodward, 1834-94, municipal engineer(B).
    PICE - 118, 462.
    PIMCE - 21, 325-326.

Parry, Harry Blackburne, 1848-83, Public Works engineer in India.
    PICE - 75, 316-317.

Parry, Rear-Adm. Sir William Edward, F.R.S., 1790-1855, hydro-
    grapher(B).
    PICE - 15, 90-92.

Parson, John, 1816-74, railway director.
    PICE - 41, 227-228.

Parsons, John Meeson, 1798-1870, railway director(B,D).
    PICE - 31, 252-253.

Parsons, Perceval Moses, 1819-92, inventor & metallurgist(B).
    Eng - 74, 418-419.
    Engng - 54, 611.
    Iron - 40, 451.
    JISI - 42, 295-296.
    PICE - 111, 385-389.

Parsons, William, 1796-1857, municipal surveyor.
    PICE - 17, 103-104.

Pasley, Maj.-Gen. Charles, 1824-90, Royal Engineer; Director of
    Works to the Admiralty(B,D).
    PICE - 103, 388-392.

Pasley, Lt.-Gen. Sir Charles William,F.R.S., 1780-1861, Royal
Engineer; Inspector-General of Railways(B,D).
PICE - 21, 545-560.

Paterson, Ebenezer Stevan, 1871-93, ---
TIESS - 36, 322-333.

Paterson, Murdoch, 1826-98, contractor.
PICE - 135, 351-353.

Paterson, Thomas, 1830-69, railway engineer.
Engng - 9, 125*.
PICE - 31, 224-225.

Paterson, William, 18(09)-81, railway engineer.
Eng - 52, 348*.
Engng - 31, 592.

Patey, Charles Henry Bennet, 1844-89, Secretary for Postal
Telegraphs(B).
Elec - 22, 623-624; 23, 239.
Elec Rev - 24, 392-393.
JIEE - 18, 282-294.

Paton, John, 1822-84, iron works manager.
JISI - 25,555.
PICE - 74, 290.

Patterson, Michael, 1830-85, paper consultant.
PICE - 83, 443-444.

Patterson, William Hammond, 1847-96, Secretary, Durham Miners'
Association.
Eng - 82, 87.

Pattinson, John, 1819-86, railway engineer in Russia.
PIME - 1886, 463.

Paul, Matthew, 18(07)-93, engineering manufacturer.
Eng - 76, 453.

Pauling, Henry John, 1821-92, rly engr in South Africa(B).
Iron - 40, 319.
PICE - 112, 359.

Pauling, Henry Richard Clarke, 1857-97, rly engr in S.Africa.
PICE - 131, 374-375.

Paulson, Webster,   -1887, civil engineer in Malta.
PICE - 91, 452-453.

Paynter, Lindley William, 1840-99, engineer in India.
PICE - 137, 432-433.

Payton, William, 18(42)-92, telegraph company secretary.
Elec - 29, 211.

Peace, Alfred Lindley, 1837-98, civil engineer.
PICE - 136, 359-360.

Peace, William Kirby, 18(22)-98, manufacturer.
Eng - 85, 209-210.
Engng - 65, 280.

Peace, William Maskell, 1834-92, Secretary, Mining Association.
Iron - 40, 427.
JISI - 41, 293-295.

Peacock, Ralph, 1826-87, manufacturing engineer.
PIME - 1869, 337-338.

Peacock, Richard, M.P., 1820-89, locomotive manufacturer(B).
Eng - 67, 207.
Engng - 47, 240*, 311-312.
Iron - 33, 210.
JISI - 34, 217-218.
PICE - 97, 404-407.
PIME - 1889, 197-199.

Peacock, William Henry, 18(55)-96, ironmaster.
TMAE - 6, 320.

Peaker, George, 1838-84, armaments engineer in India.
PIME - 1884, 474.

Pearce, Alfred W   , 1827-86, manufacturing engineer &
shipbuilder.
Engng - 41, 108.

Pearce, Richard, 1843-98, railway engineer in India.
PIME - 1898, 539-540.

Pearce, Robert Webb, 1831-89, railway engineer in India.
PIME - 1890, 292-293.

Pearce, Sir William, Bart.,M.P., 1833-88, shipbuilder(B,D).
Eng - 66, 522.
Engng - 46, 600.
Iron - 32, 548.
JISI - 34, 213-215.
TIESS - 32, 324.
TINA - 30, 466-467.

Pearse, William Monro, 1840-95, gas engineer.
PICE - 121, 337.

Pearson, Richard, 18(26)-93, ---
TMAE - 3, 286*.

Pease, Arthur, 1837-98, coal & ironmaster(B).
    Eng - 86, 240*.
    JISI - 54, 330.

Pease, Charles, 1842-73, entrepreneur.
    Engng - 16, 28*.

Pease, Edward, 1767-1858, railway engineer(B).
    Eng - 6, 103*.

Pease, Henry, 1807-81, railway director(B).
    Eng - 51, 413.
    Engng - 31, 596.
    Iron - 17, 400.

Pease, Henry Fell, 1838-96, coal owner(B).
    JISI - 51, 312.

Pease, Joseph, 1789-1872, coal owner(B).
    Engng - 13, 106.

Pease, Joseph Beaumont,   -1873, blast furnaceman.
    Engng - 16, 28*.

Pease, Walter,   -1871, manufacturer.
    Eng - 32, 408*.

Peebles, David Bruce, 1828-99, electrical manufacturer.
    Elec Rev - 45, 934.
    Eng - 88, 599.
    Engng - 68, 789.

Peel, George, 1803-87, boiler manufacturer.
    PICE - 90, 435-436.

Peel, George, Jun., 1828-75, mechanical engineer.
    PIME - 1876, 24.

Peel, William de Winton, 1850-84, railway engr in India.
    PICE - 79, 371-372.

Peet, Henry, 1813-65, railway engineer.
    PIME - 1866, 14.

Peggs, James Wallace, 1848-99, consulting civil engineer.
    PICE - 136, 360.
    TSE - 1899, 259.

Pellatt, Apsley, 1791-1863, glass manufacturer(B).
    PICE - 23, 511-512.

Pender, Sir John, 1815-96, submarine telegraph pioneer(B,D).
    Elec - 37, 334-335, 379-380.
    Elec Rev - 39, 52, 65-66.
    Engng - 62, 45, 620.

Penman, William, 1855-99, surveyor.
PICE - 139, 381-382.

Penn, John, 1770-1843, mechanical engineer.
PICE - 3, 13-14.

Penn, John, F.R.S., 1805-78, mechanical & marine engr(B,D).
Eng - 46, 229, 242; 87, 81+port.
Engng - 26, 260, 300-301.
Iron - 12, 396.
JISI - 13, 609-612.
PICE - 59, 298-302.
PIME - 1879, 13-15.
TINA - 20, 263-264; 21, 266.

Penny, Alfred, 1811-90, gas engineer.
PICE - 101, 300-301.

Penny, Alfred, Jun., 1841-91, Public Works engr in India.
PICE - 97, 406-407.

Penson, Thomas, 1839-60, architect & surveyor.
PICE - 20, 156*.

Pentland, Augustus Tichborne, 1857-1900, Public Works engineer
in Ireland.
PICE - 143, 320-321.

Percy, John, M.D., F.R.S., 1817-89, metallurgist(B,D).
Eng - 67, 551.
Engng - 47, 732.
Iron - 33, 577.
JISI - 34, 210-213.
PICE - 99, 343-346.

Perkins, Loftus, 1834-91, inventor of high pressure steam
engine(B,D).
Eng - 71, 349.
PICE - 105, 311-314.
PIME - 1891, 192-193.

Perring, John Shae, 1813-69, railway plant manufacturer(B).
PICE - 30, 455-456.
PIME - 1870, 15-16.

Perry, Alfred, 1834-92, lighthouse designer.
PICE - 112, 371.
PIME - 1892, 407-408.

Perry, F   C   , 18(18)-1900, engineering manufacturer.
Eng - 89, 264.

Perry, J   ,   -1879, railway director & manufacturer.
Engng - 27, 294*.

Perry, Thomas Joseph, 1824-85, ironfounder.
    Iron - 25, 297.
    JISI - 27, 539.
    PIME - 1885, 301-302.

Peto, Sir Samuel Morton, 1809-89, contractor(B,D).
    Eng - 68, 438.
    Engng - 48, 634.
    Iron - 34, 445.
    PICE - 99, 400-403.

Pfeil, Frederick Molesworth, 1835-73, civil engr in S.Africa.
    PICE - 38, 320-321.

Philipson, John, 1832-98, coach manufacturer.
    Eng - 86, 22.
    PIME - 1898, 316-317.

Phillips, Alfred, 1844-89, municipal engineer(B).
    PICE - 97, 422.

Phillips, Charles George Washington, 1856-1900, railway engr.
    PICE - 143, 335.

Phillips, David, 1831-94, marine engineer(B).
    PICE - 118, 450-452.

Phillips, Exham, 1847-99, iron works manager.
    PIME - 1899, 141.

Phillips, Henry Parnham, 1856-96, railway engr in India.
    PICE - 125, 418-419.

Phillips, John, 18(17)-97, sanitary engineer.
    Engng - 64, 203.

Phillips, John Arthur, F.R.S., 1822-87, consultant mining
    engineer; metallurgist(B,D).
    Elec Rev - 20, 57.
    Iron - 29, 31.
    PICE - 89, 481-484.

Phillips, Samuel Elkins, 18(47)-93, telegraph engineer.
    Elec - 31, 341.
    Elec Rev - 32, 656-657; 33, 99.

Phipps, George Henry, 1807-88, railway engineer(B).
    PICE - 96, 330-333.

Phipps, George Henry, Jun., 1840-84, railway engr overseas.
    PICE - 78, 439-441.

Phipson, Wilson Weatherley, 1838-91, heating engineer(B).
    PICE - 108, 406-408.

Pickles, George Herbert, 18(76)-1900, telephone engineer.
Elec Rev - 46, 885.

Pidgeon, Daniel, 1833-1900, agricultural implement manufr.
JISI - 58, 391.
PICE - 142, 395-396.

Piercy, Benjamin, 1827-88, railway engineer(B).
Eng - 65, 283.
Engng - 45, 334-359.
Iron - 31, 277.
JIEE - 96, 333-339.

Piercy, Robert, 1825-94, railway contractor.
PICE - 118, 452-453.

Pigott, George, 1833-71, inventor & manufacturer.
PIME - 1872, 20-21.

Pilbrow, James, 1813-94, drainage engineer.
Eng - 77, 175.
PICE - 117, 381-382.

Pinchin, Robert, 1821-88, surveyor in South Africa(B).
PICE - 95, 388-391.

Piper, Joseph, 18(07)-83, iron works manager.
JISI - 23, 671-672.

Piper, William, 1818-1900, contractor.
PICE - 140, 293.

Pirrie, John Sinclair, 1849-96, mechanical engr & manufacturer.
PICE - 125, 419.
PIME - 1896, 97.

Pitts, George Albert, 1849-82, consulting engineer.
PIME - 1883, 24.

Pitts, Joseph, 1812-70, company representative.
PICE - 31, 253.
PIME - 1871, 16-17.

Platt, James, 1834-97, consulting & manufacturing engineer.
Elec Rev - 41, 934*.
Eng - 84, 653-654.
Engng - 64, 797.
PICE - 132, 387-388.
PIME - 1897, 516-517.

Platt, John, M.P., 1817-72, engineering manufacturer(B).
PIME - 1873, 19-20.

Platt, John, 1834-97, engineering manufacturer.
    JISI - 53, 319.

Player, John, 1808-70, mechanical engineer(B).
    PIME - 1871, 17-18.

Plews, John, 1795-1861, dock engineer.
    PICE - 21, 564-567.

Plowright, Robert, 1838-98, mining engineer.
    TFIME - 16, 126.

Plum, Thomas William, 1814-78, iron works manager.
    PIME - 1879, 15.

Plummer, E   ,   -1898, ---
    Eng - 85, 23*.

Pochin, Henry Davis, 1824-95, iron & coal merchant(B).
    Engng - 60, 542.
    JISI - 48, 352-353.

Poke, George Henry, 1845-91, armaments engineer.
    PIME - 1893, 95.

Pole, William, F.R.S., 1814-1900, Professor of Civil Engineering,
    University College, London(B,D).
    Elec Rev - 48, 26.
    PICE - 143, 301-309.
    TSE - 1900, 270.

Pollard, Thomas Sydney, 1859-89, railway engineer.
    PICE - 99, 377.

Pollock, William, 1864-95, railway engineer.
    PICE - 124, 432-433.

Pontifex, Samuel, 1816-86, consultant gas & water engineer.
    PICE - 85, 408.

Poole, Braithwaite, 1805-88, Chairman, Railway Clearing House.
    Eng - 66, 115.

Poole, James,   -1872, railway chairman.
    Engng - 15, 15*.

Pooley, Henry, 1803-78, weighing machine manufacturer(B).
    Eng - 46, 175.
    PICE - 55, 331-334.

Pope, John, 1820-47, Colonial Engineer in Hong Kong.
    PICE - 8, 18.

Porter, John, 1860-1900, mechanical engr in India.
    PICE - 143, 335-336.

Porter, John Francis, 1810-65, consulting engineer.
    PICE - 26, 582-583.

Porter, John Henderson, 1824-95, cast-iron manufacturer.
    PICE - 124, 441-443.

Pothecary, George, 1841-76, engineer in India.
    Engng - 22, 91*.
    PICE - 45, 257-259.

Potter,   , 18(27)-80, ironmaster.
    Iron - 15, 155.

Potter, James, 1801-57, railway engineer.
    PICE - 17, 94-96.

Potter, William Auboné, 1832-87, mining engineer(B).
    PICE - 91, 421-423.

Potts, Arthur, 1814-88, locomotive manufacturer.
    PICE - 96, 339-340.

Potts, Cuthbert,   -1879, consulting engineer.
    Eng - 83, 451*.

Potts, Dr Lawrence Holker, 17(90)-1850, inventor & mech. engr.
    CEAJ - 13, 144.

Powell, Thomas, 1780-1863, colliery owner.
    Eng - 15, 187*.

Powell, William, 1824-82, civil engineer.
    PICE - 71, 414-415.
    PIME - 1883, 25.

Power, Samuel, 1814-71, civil engineer.
    PICE - 33, 236-241.

Pratt, Mervyn James Butler, 1857-84, civil engineer.
    PICE - 79, 372.

Preece, George Edward, 1838-95, electrical engineer.
    Elec - 34, 437.
    Elec Rev - 36, 166-167, 197.

Preston, Francis, 1823-91, mechanical engineer.
    PIME - 1892, 99*.

Preston, Frederick Charles, 1867-94, mechanical engineer.
    PICE - 123, 444-445.

Preston, Robert Berthon, 1820-60, mechanical engineer(B).
    PICE - 20, 157-158.

Price, Dr Astley Paston, 1826-86, industrial chemist(B).
    PICE - 87, 458-460.

Price, Edward, 18(05)-71, railway engineer.
    Eng - 31, 248.
    PICE - 33, 267-269.

Price, Edward Bellingham, 1859-98, Public Works engineer in
    Australia.
    PICE - 133, 404-405.

Price, James, 1831-95, civil & railway engineer(B).
    PICE - 121, 327-329.

Price, John, 18(34)-1900, colliery manager.
    Eng - 89, 498.

Price, Samuel Thomas, 1847-93, gas works manager.
    TSE - 1893, 237*.

Price, William, 1826-83, iron works manager.
    JISI - 23, 667.

Prickett, Lancelot George, 1856-95, Public Works engineer in
    India(B).
    PICE - 122, 399-400.

Priestley, Alfred Coveney, 1837-95, railway engineer.
    PICE - 121, 338.

Priestman, Jonathan, -1888, coal & ironmaster.
    Eng - 66, 543.

Prime, Charles, 1834-90, engineer in Ceylon.
    PICE - 102, 330-331.

Prime, T , -1892, electroplate manufacturer.
    Elec - 29, 109.

Prince, Paul, 18(50)-95, railway signal engineer.
    Elec Rev - 37, 391.

Pringle, Thomas, -1897, mining engineer.
    Eng - 83, 425*.

Prior, Jonathan Charles, -1849, coke manufacturer.
    PICE - 9, 104.

Pritchard, Daniel Baddeley, 1827-72, consulting engineer in
    Australia.
    PICE - 38, 293-295.

Pritchard, Edward, 1838-1900, consulting engineer.
    Eng - 89, 540.
    Engng - 69, 660.
    PICE - 141, 348-349.
    PIMCE - 26, 250-252. Port. frontis. v.19.

Pritchard, W   ,   -1898, ---
    Eng - 85, 23*.

Proctor-Sims, Richard,   -1900, Public Works engr in India.
    PICE - 142, 372-373.

Prosser, William Henry,1843-94, marine engineer.
    Eng - 77, 153.
    PIME - 1894, 598-599.

Protheror, David Thomas Rhys, 1856-96, sanitary engineer.
    PICE - 125, 419-420.

Provis, William Alexander, 1792-1870, railway engineer(B).
    PICE - 31, 225-230.

Purchas, Samuel Guyon, 1823-94, municipal surveyor.
    PICE - 117, 382-383.

Purdon, Frederick, 1853-97, civil engineering contractor.
    PICE - 129, 395-396.

Purdon, Wellington, 1815-89, railway engineer.
    PICE - 97, 408-413.

Purkiss, Henry John, 18(42)-65, naval architect & marine engr(B).
    TINA - 6, xxiv.

Purnell, Edward James, 1856-98, municipal engineer.
    PICE - 132, 397.

Putman, William, 1835-97, forge master.
    Eng - 83, 461.
    JISI - 52, 258.
    NECIES - 13, 272.
    PIME - 1897, 138-139.

Quick, Edward, 1857-95, consulting engineer.
    PICE - 124, 433-434.

Quick, Joseph, 1809-94, water engineer.
    PICE - 117, 383-384.

Quirk, Edward Philpot Senhouse, 1850-97, mechanical engineer.
    PIME - 1897, 139-140.

Radcliffe, James, 1828-92, telegraph superintendent.
    Elec - 29, 161.
    Elec Rev - 30, 471.

Radcliffe, John Alexander, 1823-91, railway solicitor.
    PICE - 104, 320-321.

Radford, William, 1816-54, civil engineer.
    PICE - 14, 136-137.

Radford, William, 1817-97, consulting engineer & surveyor.
    PICE - 131, 376-377.

Raine, Nicholas, 1826-98, iron works manager.
    Eng - 85, 243.

Rake, Alfred Stansfield, 1831-71, consulting naval architect.
    PICE - 33, 270.
    PIME - 1872, 21.

Rammell, Charles, 1822-56, railway contractor.
    Eng - 2, 476.
    PICE - 16, 127-128.

Ramsay, George Heppel, 17(90)-1879, colliery owner.
    Iron - 14, 714.

Ramsbottom, John, 1814-97, mechanical & railway engineer(B).
    Eng - 83, 549, 568+port.
    Engng - 63, 722, 751-752.
    JISI - 51, 312-313.
    PICE - 129, 382-385.
    PIME - 1897, 236-241.
    TMAE - 7, 283-287.

Ramsden, Sir James, 1822-96, rly developer & steel manufr(B).
    Eng - 82, 421.
    Engng - 62, 53 .
    JISI - 50, 260-261.
    PICE - 129, 385-389.
    TINA - 38, 313.

Randall, William Sandcroft, 1851-1900, gas engineer.
    PICE - 143, 336.

Randolph, Charles, 1809-78, shipbuilder & marine engineer(B).
    Engng - 26, 398*, 419-420.
    Iron - 12, 622.
    TIESS - 33, 3.

Rankin, Daniel, 1824-85, manufacturing engineer.
    Engng - 40, 469.
    Iron - 26, 438.
    TIESS - 29, 221.

Rankine, William John Macquorn, F.R.S., 1820-72, Professor of
    Engng & Mechanical Science, Glasgow University(B,D).
    Eng - 34, 434; 35, 1.
    Engng - 14, 437; 15, 13-15.
    PIME - 1873, 21.
    TINA - 14, 234-237.

Ransome, Frederick, 1818-93, cement manufacturer(B).
    Eng - 75, 412.
    Engng - 55, 624.
    Iron - 41, 385.
    JISI - 43, 173*.
    PICE - 115, 402-404.

Ransome, James, 1782-1850, agricultural implement manufr.
    PICE - 10, 99-100.

Ransome, James Allen, 1806-75, agricultural implement manufr(B).
    Eng - 39, 334*.
    PICE - 41, 228-232.

Ramsome, Robert, 1795-1864, agricultural implement manufr(B).
    PICE - 25, 529-531.

Ransome, Robert Charles, 1830-86, agricultural machinery manufr(B).
    Eng - 61, 213.
    Engng - 41, 258.
    Iron - 27, 229.
    JISI - 29, 803-804.
    PIME - 1886,122-123.

Ransome, Robert James, 1830-91, railway plant manufacturer.
    JISI - 39, 243-245.
    PIME - 1891 292.

Ranson, Robert Gill, 1792-1843, paper manufacturer.
PICE - 3, 15.

Rapier, Richard Christopher, 1836-97, railway plant manufr(B).
Eng - 83, 573+port.
Engng - 63, 754.
JISI - 51, 313.
PICE - 129, 389-391.
PIME - 1897, 140-141.

Rastrick, John Urpeth, 1780-1856, mechanical engr & manufr(B).
Eng - 2, 627.
PICE - 16, 128-133.

Raven, Percy Earle, 18(58)-1900, engineer in India.
Eng - 89, 618*.

Ravenhill, John Richard, 1824-94, marine engine manufacturer(B).
PICE - 120, 364-365.
PIME - 1895, 145-146.

Ravenhill, Richard, 1800-87, marine engine manufacturer(B).
PICE - 89, 495-496.

Rawlins, John, 1836-98, carriage & wagon works manager.
PICE - 133, 415-416.
PIME - 1898, 317-318.

Rawlinson, Sir Robert, 1810-98, Chief Engineering Inspector,
Local Government Board(B,D).
Elec Rev - 42, 774*.
Eng - 85, 528.
Engng - 65, 699.
PICE - 134, 386-391. Port. frontis. v.119.
PIMCE - 24, 365-366.
TSE - 1898, 237-238.

Rawlinson, William,   -1874, engineer in Brazil.
PICE - 43, 318.

Rawnsley, Henry Charles,   -1852, ---
PICE - 12, 168*.

Rayne, Middelton, 1830-82, Public Works engr in India.
PICE - 71, 415-416.

Rayner, Thomas Cheveley, 1840-98, contractor.
PICE - 135, 368-369.

Read, Samuel, 17(95)-1863, naval architect(B).
Eng - 16, 174*.
TINA - 4, xxi-xxii.

Read, Thomas Croad, 1854-95, Lloyd's surveyor(B).

Eng - 79, 177.
Engng - 59, 276.
TINA - 35, 362-363.

Reah, Hudson, 1839-98, municipal engineer.
PICE - 134, 415-416.

Reckenzaun, Anthony, 1850-93, electrical engineer(B).
Elec - 32, 66.
Elec Rev - 33, 532-533.
Eng - 76, 468.
Engng - 56, 615, 698.

Redhead, S   ,   -1900, colliery owner.
Eng - 89, 186*.

Redl, Charles Arthur, 1813-94, signal engineer(B).
Eng - 78, 133.

Redman, John Baldry, 1816-99, consulting civil engineer(B).
PICE - 142, 373-376.

Reeks, Trenham, 18(23)-79, Registrar, Royal School of Mines(B).
Eng - 47, 366.

Reid, Alfred George Woodward, 1849-94, engineer in India.
PICE - 121, 329-330.

Reid, David, 1841-92, railway engineer overseas(B).
PICE - 110, 396-398.

Reid, Henry David Alexander, 1856-99, rly engr in Latin
America.
PICE - 136, 355-357.

Reid, James, 1822-94, locomotive manufacturer(B).
Eng - 77, 570.
Engng - 57, 846.
PICE - 117, 385-387.
PIME - 1894, 278-279.
TIESS - 37, 196-197.

Reid, John, 18(16)-93, shipbuilder.
Eng - 76, 225.

Reid, Robert Carstairs, 1845-94, water engineer(B).
PICE - 116, 369-372.

Reilly, Callcott, 1829-1900, Professor of Engineering Constr-
uction, Royal Engng College, Coopers Hill.
Engng - 69, 717.
PICE - 142, 376-379.

Remfrey, Charles, 1837-92, mining engineer in Spain.
PICE - 111, 401-402.

Remington, George,   -1883, railway contractor.
Iron - 22, 270.

Rendel, James Meadows, F.R.S., 1799-1856, harbour engr(B,D).
CEAJ - 19, 402-403.
Eng - 2, 650.
PICE - 16, 133-142.

Rendel, William Edgecumbe, 1820-81, inventor.
Eng - 52, 182.

Rendel, William Stuart, 18(56)-98, ---
JISI - 53, 320*.

Rennie, George, F.R.S., 1791-1868, contracting engineer(B,D).
Eng - 21, 252*.
PICE - 28, 610-615.

Rennie, Sir John, F.R.S., 1794-1874, civil engineer(B,D).
Eng - 38, 209.
Engng - 18, 206-207.
PICE - 39, 273-278.
Prac Mag - 4, 321-325.

Rennie, William Coupar, 1843-91, Public Works engr in India(B).
PICE - 105, 314-316.

Renton, Henry, 1815-51, consulting civil engineer.
PICE - 11, 105-106.

Reynolds, Edward, 1825-95, engineering manufacturer.
Eng - 79, 59.
NECIES - 12, 245-246.
PICE - 121, 331-332.
PIME - 1895, 146-147.

Reynolds, Frederick Cornell, 1836-68, civil engineer.
PICE - 30, 475-476.
PIME - 1869, 17.

Reynolds, John, 1796-1847, ironmaster.
PICE - 7, 12-14.

Reynolds, Robert, 1833-99, railway engineer overseas.
PICE - 138, 491-494.

Rhind, James, 1848-88, mech. & rly engineer in India(B).
PICE - 99, 377-379.

Rhodes, Alexander, 18(36)-77, Public Works engr in India.
PICE - 50, 187.

Rhodes, Thomas, 1789-1868, railway & dock engineer(B).
PICE - 28, 615-618.

Rhodes, William, 1812 -85, Caledonian Canal Engineer.
  PICE - 83, 437-438.

Rich, William Edmund, 1844-86, engineering manufacturer.
  Elec Rev - 20, 40.
  Eng - 63, 9.
  Engng - 43, 8-10.
  Iron - 29, 31.
  PICE - 89, 484.
  PIME - 1887, 148-149.

Richards, Evan Matthew, 1820-80, steel manufr & coal owner(B).
  JISI - 17, 688-689.

Richards, Admiral Sir George Henry, 1819-96, hydrographer &
  telegraph engineer(B).
  Elec - 38, 120.

Richards, Josiah, 1823-88, iron & tinplate manufacturer.
  PIME - 1888, 156.

Richards, Josiah J   ,   -1894, steel works manager.
  JISI - 45, 393*.

Richards, Lewis, 1829-98, iron & steel works engineer.
  JISI - 53, 320-322.
  PIME - 1898, 540-542.

Richardson, Capt. A   H   J   , 18(41)-87, marine engr.
  Elec - 19, 415.

Richardson, Charles, 1814-96, civil engineer(B).
  Engng - 61, 219.
  PICE - 123, 417-419.

Richardson, George, 1845-84, textile machine manufacturer.
  JISI - 25, 557.
  PIME - 1884, 474-475.

Richardson, Henry Yarker, 1819-70, entrepreneur.
  PICE - 33, 270-271.

Richardson, Joshua, 1799-1886, mining engineer(B).
  PICE - 86, 358-363.

Richardson, Robert, 1812-91, consulting civil engineer(B).
  PICE - 97, 407-409.

Richardson, Thomas, F.R.S., 1816-67, manufacturing chemist(B,D).
  PICE - 30, 476.

Richardson, Thomas, M.P., 1821-91, marine engr & ironmaster(B).
  Engng - 51, 21-22.
  JISI - 39, 245-246.
  NECIES - 7, 1.

Richardson, William, 17(96)-1891, gas & water engineer.
Engng - 51, 587.

Richardson, William, 1811-92, textile machinery manufr(B).
Eng - 75, 9.
Engng - 54, 782; 55, 11.
JISI - 41, 288-289.
PIME - 1892, 408-410.
TMAE - 2, 280-281.

Riches, J   Osborne,   -1886, colliery owner.
Engng - 42, 10*.

Rickard, Percy, 1859-93, railway engineer.
PICE - 115, 393-394.

Rickard, William,   -1898, cable manufacturer.
Elec Rev - 41, 891*.

Ricketts, Frederick Henry, 1837-74, telegraph engineer.
PIME - 1875, 30.

Rickman, John,   -1840, statistician.
CEAJ - 4, 127-218.
PICE - 1, 15-17.

Rickman, William Charles, 1812-86, architect & amateur engr(B).
PICE - 86, 374-376.

Riddell, Robert, 1840-90, railway engineer in India(B).
PICE - 106, 337-341.

Ridehalgh, George John Miller, 1835-92, amateur mechanical
engineer.
PIME - 1892, 410.

Rider, Henry Hyam, 18(57)-1900, wagon manufacturer.
JISI - 57, 256.

Ridley, John, 1806-87, inventor of mechanical reaper.
Iron - 30, 530.

Ridley, John Hindmarsh, 1858-99, contractor.
Eng - 87, 552.

Ridley, Samuel C   ,   -1891, railway contractor.
Iron - 38, 541.

Ridley, Thomas Dawson, 1825-98, civil engineer.
Eng - 85, 56.
JISI - 53, 323.
PICE - 131, 394-395.

Rigden, John Lambe, 1858-91, rly engineer in Natal.
PICE - 109 423-424.

Riley, George Farrer, 18(48)-99, boilermaker.
Eng - 87, 43.

Riley, John, 18(09)-97, ironmaster.
Eng - 84, 223*.

Ring, Robert, 1846-90, engineer in Burma(B).
PICE - 100, 401-403.

Ritchie, Joseph Horatio, 1799-1872, naval architect.
TINA - 13, 331-334.

Ritchie, Robert, 1795-1871, ventilating & heating engr(B).
PICE - 33, 271-272.

Robe, Lt.-Col. Alexander Watt, 1793-1849, Royal Engineer.
CEAJ - 13, 20-21.
PICE - 9, 105-106.

Roberts, Charles Warren, 1852-97, railway engineer.
PICE - 129, 396.

Roberts, Griffith, 1868-99, municipal engineer.
PICE - 138, 497.

Roberts, Llewellyn, 18(25)-98, iron & steel manufacturer.
Eng - 86, 22.

Roberts, Peter, 1846-88, mechanical engineer.
PICE - 96, 349-350.

Roberts, Richard, 1789-1864, inventor & manufacturer(B,D).
CEAJ - 27, 146-147.
Eng - 17, 183.
PICE - 24, 536-539.
TINA - 5, xxvi.

Roberts, Robert William, 1830-93, quarry manager.
PICE - 113, 376.

Roberts, Samuel, 1800-87, steel & waterworks director.
Iron - 30, 530.

Roberts, Samuel Ussher, 1821-1900, Public Works engr in Ireland.
PICE - 140, 279-281.

Roberts, Thomas, 1826-96, railway engineer in India.
TSE - 1897, 201-202.

Roberts, Thomas, 1837-1900, consulting civil engineer.
PICE - 140, 283-284.

Robertson, George, 1830-96, civil engineer.
PICE - 124, 419-420.

Robertson, Henry, 1816-88, rly contractor & ironmaster(B).
Eng - 65, 283.
Engng - 45, 359.
Iron - 31, 277.
JISI - 33, 219-220.
PICE - 93, 489-492.
PIME - 1888, 264-265.

Robertson, James, 1818-89, chief railway engineer(B).
PICE - 99, 355-364.

Robertson, James, 18(19)-97, inventor.
Eng - 88, 599.
Engng - 68, 757.

Robertson, James, 1835-76, railway engineer.
PICE - 44, 234-235.

Robertson, John,   -1872, shipbuilder.
Engng - 14, 365.

Robertson, Robert Andrew, 1843-1900, sugar refiner.
PICE - 142, 379.
TIESS - 43, 363-364.

Robertson, William, 1833-98, mechanical engineer.
PIME - 1899, 141-142.

Robinson, Maj.-Gen. Daniel George, 1826-77, Bengal Engineer(B).
JSTE - 6, 492-496.

Robinson, James, 1852-96, municipal surveyor.
PICE - 128, 361-362.
PIMCE - 23, 479-480.

Robinson, James Salkeld, 1849-92, machinery manufacturer.
Eng - 74, 75*.
Engng - 54, 109*.
Iron - 40, 78.
PIME - 1892, 568.

Robinson, John, 1823-77, machinery manufacturer(B).
PIME - 1878, 13-14.

Robinson, Joseph, 1818-83, ironmaster.
PICE - 76, 372.

Robinson, Murrell Robinson, 1821-1900, Colonial Engineer.
PICE - 140, 281.

Robinson, Admiral Sir Robert Spencer, F.R.S., 1809-89, naval
architect(B,D).
TINA - 31, 288.

Robinson, Stephen, 1794-1881, mining engineer(B).
    Iron - 18, 424.
    PICE - 68, 312-314.

Robinson, William, 1819-88, admiralty engineer.
    Iron - 31, 55.

Robson, Edward, 1830-94, iron ore merchant.
    JISI - 46, 265*.

Robson, Neil, 1807-69, civil & mining engineer(B).
    PICE - 30, 456-457.

Robson, Richard Oswald, 1826-93, surveyor in Victoria.
    PICE - 114, 384-385.

Robson, Robert, 1817-72, iron founder & mechanical engr.
    PIME - 1873, 17-18.

Robson, Thomas, 1836-90, mining engineer.
    PIME - 1890, 173-174.

Roden, William Sargeant, 1829-82, ironmaster(B).
    JISI - 21, 654-655.

Rodger, James, 1814-73, shipbuilder & engineer.
    Engng - 15, 362.

Rodger, William, 1843-93, consulting engr in India.
    PIME - 1893, 498.

Rodgers, Maurice George, 1857-98, manufacturer.
    Eng - 85, 504.
    Engng - 65, 663.

Roe, John, 1795-1874, sanitary engineer(B).
    PICE - 39, 297-298.

Roe, John Phanuel, 1814-88, iron works engineer(B).
    JISI - 34, 222-224.
    PICE - 98, 405-408.

Roebuck, W   , 18(29)-92, railway engineer.
    Iron - 40, 276.

Roehricht, Richard Hugo Oswald, 1834-90, railway draughtsman.
    PICE - 106, 346-347.

Rogers, Archibald Colin Campbell, 1849-92, civil engr in India.
    PICE - 113, 355-356.

Rogers, Charles Henry, 1854-94, Public Works engr in Trinidad.
    PICE - 119, 408-409.

Rogers, W   R   ,   -1892, works manager.
    Iron - 39, 517.

Rogers, William, 1843-88, railway engineer overseas.
    PICE - 95, 380-382.

Rogerson, John, 1828-94, iron works manager.
    JISI - 45, 393-394.
    PICE - 117, 399-400.

Rollo, David, 1820-90, marine engine manufacturer.
    Engng - 49, 673.
    PIME - 1890, 293.

Rolt, Peter, 1798-1882, ironmaster(B).
    Iron - 20, 212.

Ronalds, Sir Francis, F.R.S., 1788-1873, electrical engr(B,D).
    Eng - 36, 107.
    JSTE - 9, 196-198.

Ronayne,Joseph Philip, 1822-76, contractor & surveyor(B,D).
    PICE - 46, 274-276.
    PIME - 1877, 21-22.

Roper, Richard Steven, 1835-94, iron works secretary.
    PICE - 45, 259-260.

Roper, Robert, 1854-94, company secretary.
    JISI - 45, 393*.

Roscoe, Thomas Ridyard,   -1896, municipal engineer.
    PIMCE - 23, 480*.

Rose, Thomas, 1809-74, ironmaster.
    PIME - 1875, 31.

Ross, Alexander McKenzie, 18(06)-62, civil engineer.
    Eng - 14, 93*, 193.

Ross, Charles, 18(41)-99, engineering manufacturer.
    Engng - 68, 692.

Ross, George Hennet, 1843-88, railway engr overseas.
    PICE - 95, 382-383.

Ross, Henry Francis, 1819-94, civil engineer in Spain(B).
    PICE - 116, 384-386.

Ross, John, 1812-61, railway carriage works manager.
    PIME - 1862, 17-18.

Ross, Leathom Earle, 1837-86, contracting civil engr(B).
    PICE - 89, 492-494.

Ross, Owen Charles Dalhousie, 1823-95, civil engr & inventor(B).
    PICE - 122, 382-385.

Ross, Thomas Braban, 1842-78, mining engineer.
    PIME - 1879, 15.
    TCMCIE - 7, 12.

Rosser, William, 1829-94, surveyor & engineer(B).
    PICE - 121, 341-343.

Rothwell, Edmund, 1858-98, municipal engineer.
    PICE - 132, 397.
    PIMCE - 24, 366.

Rothwell, Peter, 1792-1849, steam engine manufacturer.
    PICE - 9, 100-101.

Rotton, Henry Beadon, 1836-71, railway engineer.
    PICE - 33, 273.

Round, Benjamin, 1836-1900, pig-iron manufacturer.
    JISI - 58, 391-392.

Routledge, Thomas, 1819-87, paper manufacturer(B).
    PICE - 92, 404-406.
    PIME - 1887, 469-470.

Rowan, David, 1822-98, marine engineer.
    Eng - 86, 135.
    PICE - 134, 407-408.
    TIESS - 41, 380-381+frontis.

Rowan, Capt. Frederick Charles, 1844-93, consulting engineer in
    Melbourne.
    Elec - 30, 350.
    PICE - 113, 377.

Rowan, W R   ,   -1900, tramway engineer.
    Eng - 90, 243+port.

Rowan, William,   -1884, marine engineer.
    Eng - 57, 57.

Rowe, Richard Reynolds, 1824-99, municipal engineer.
    PICE - 141, 349-350.

Rowles, Henry,   -1840, ironmaster.
    CEAJ - 4, 127.
    PICE - 1, 15.

Rowley, Samuel Fellows, 18(60)-88, municipal engineer.
    PIMCE - 14, 400.

Roy, Norman William, 1862-96, railway engr in India.
    PICE - 126, 405-406.

Ruglen, James, 18(21)-93, mining journalist.
    Eng - 76, 577.

Rumball, Alfred, 1831-96, railway engr in South America.
    PICE - 128, 356-358.

Rumble, Thomas William, 1832-83, security expert & civil engr(B).
    PICE - 73, 368-369.
    PIME - 1884, 67-68.

Rummens, Francis, 1810-92, railway engineer.
    JISI - 44, 292.

Rundle, Cubitt Sparkhall, 1818-89, Public Works engr in India.
    PICE - 100, 403-405.

Rushton, James Raine, 1833-73, engineer in India.
    PICE - 39, 278-281.

Russ, William, 1848-91, sanitary engineer.
    PICE - 104, 295-296.

Russell, Edward,   -1876, engineering manufacturer.
    Eng - 41, 13.

Russell, George,   -1877, ironmaster.
    JISI - 11, 540.

Russell, John Scott, F.R.S., 1808-82, naval architect(B,D).
    Eng - 53, 430.
    Engng - 33, 583.
    Iron - 19, 472.
    PICE - 87, 427-440.
    TINA - 23, 258-261.

Russell, Macnamara, 1836-99, road engineer in Australia.
    PICE - 139, 367.

Russell, R   ,   -1900, civil engineer.
    Engng - 69, 662.

Russell, Thomas, 1821-86, shipbuilder.
    Engng - 42, 107.
    TIESS - 29, 221.

Ruston, Joseph, 1835- 97, engineering manufacturer.
    Eng - 83, 623.
    Engng - 63, 836.
    PIME - 1897, 141.

Ruthven, Morris West, 18(50)-96, draughtsman.
NECIES - 12, 253。

Ryan, Jeremiah, 1839-91, rly contractor in India.
PICE - 107, 422.

Ryan, Thomas Tenison, 1835-76, railway engr overseas.
PICE - 48, 272-273.

Ryder, George, 1839-92, cotton machinery manufacturer.
PIME - 1892, 569.

Ryder, William, 1808-67, inventor; machinery manufacturer.
PIME - 1868, 17-18.

Ryland, Frederick, 1845-99, engineering manufacturer.
JISI - 55, 267.
PIME - 1899, 142.

Ryland, William,   -1900, manufacturer.
JISI - 58, 392*.

Rylands, Peter, M.P., 1820-87, ironmaster(B,D).
Iron - 29, 120.

Rymer-Jones, Alexander Manson, 1845-81, railway engr overseas.
PICE - 67, 407-408.

Rymer-Jones, Thomas Ernest Manson, 1839-94, rly engr overseas(B).
Elec Rev - 38, 148.
PICE - 118, 453-455.
TSE - 1894, 255.

Sabine, Robert Henry, 1837-84, consulting telegraph engr.
Elec Rev - 25, 422*.
Eng - 38, 436.
Iron - 24, 405*.
JIEE - 13, 495.

Sach, William Henry, 18(29)-89, rly telegraph engr in India.
Elec - 23, 595.

Sackings, J J   , 1839-99, marine engineer.
Eng - 87, 40.

Sacré, Alfred Louis, 1841-97, engineering manufacturer.
    PICE - 130, 319-320.
    PIME - 1897, 141-142.

Sacré, Charles Reboul, 1831-89, railway engineer.
    PICE - 98, 399-400.

Sacré, Edward Antoine, 1838-81, locomotive manufacturer.
    PICE - 67, 419-420.

Sainsbury, Francis Charles Barrett, 1856-98, mechanical engr.
    PIME - 1898, 706.

St. George, A   R   ,   -1896, electrical engr & chemist.
    Elec Rev - 38, 280.

Salkeld, William, 18(12)-98, railway engineer.
    Eng - 86, 578.

Salmond, John Mitchell, 1842-90, Public Works engr in India.
    PICE - 102, 335.

Salter, Frank, 1848-88, engineering manufacturer.
    PICE - 96, 340-341.
    PIME - 1889, 199-200.

Samuda, Joseph D'Aguilar, 1813-85, shipbuilder & marine engr(B,D).
    Eng - 59, 338;   87, 82+port.
    Iron - 25, 384.
    PICE - 81, 334-337.
    PIME - 1885, 302.
    TINA - 25, 307-309.

Samuel, James, 1824-74, civil & railway engineer.
    Eng - 37, 364*.
    Engng - 17, 454.
    PICE - 39, 280-282.
    PIME - 1875, 31-32.

Samuelson, Alexander, 1826-73, consulting mechanical engr.
    PICE - 38, 295-296.
    PIME - 1874, 24.

Sanders, Richard Barnsley, 1845-1900, municipal surveyor.
    PICE - 142, 380.

Sanderson, Charles, 1824-70, railway engineer.
    PICE - 31, 230-233.

Sanderson, Francis, 18(50)-99, iron & steel manufacturer.
    Eng - 88, 433.

Sanderson, George, 18(08)-93, manufacturer.
    Iron - 41, 490.

Sandham, Henry, 1832-92, mechanical engineer.
PIME - 1893, 95-96.

Sang, John, 1809-87, civil engineer.
Engng - 43, 369.

Sartoris, Herbert, 1845-1900, coal & ironmaster.
JISI - 58, 392.

Saunders, George Henry, 1820-57, marine engr & manufacturer.
PICE - 17, 105-106.

Saunders, James, 18(32)-1900, iron trade representative.
JISI - 57, 256.

Saunders, John,   -1881, ironmaster.
Iron - 18, 71.

Saunders, Radcliffe,   -1898, telegraph engineer.
Elec - 40, 646.

Saunders, Richard James Harris, 1835-92, harbour engr in Brazil.
PICE - 111, 389-391.

Savill, Robert, 1807-88, railway company secretary(B).
Eng - 66, 333.
Iron - 32, 353.

Saxon, George, 1821-79, mechanical engineer.
PIME - 1880, 8.

Scaldwell, Robert Thomas, 1844-98, Public Works engr in India.
PICE - 137, 434-435.

Scamp, William, 1801-72, admiralty engineer(B).
PICE - 36, 273-278.

Schmid, Alfred Ewald, 1831-80, railway engineer.
PICE - 61, 295-296.

Schneider, Henry William, 1817-87, iron & steel manufacturer(B).
Iron - 30, 462.
JISI - 28, 220-222.
PICE - 92, 406-410.

Schofield, Christopher James, 1832-92, ironmaster, coal owner
& manufacturer.
PIME - 1892, 569-570.

Schram, John Richard, 1834-98, consulting civil engineer.
PICE - 133, 405-406.
PIME - 1898, 318.

Schriven, William Harry, 1850-98, railway engr in India.
PICE - 132, 397-398.

Scorgie, James, 1835-95, Prof. of Civil Engng, Poona。
    Engng - 60, 89.
    PIME - 1895, 541-542.
    TIESS - 38, 336-337.
    TSE - 1895, 270-271.

Scotland, John Parry, 1852-94, Public Works engr in India.
    PICE - 118, 463-464.

Scott, Maj.-Gen. Alexander de Courcy, 1834-99, Royal Engr.
    PICE - 139, 383.

Scott, Charles William, 1831-63, marine engineer.
    PICE - 23, 512-513。

Scott, Clement Augustus, 1828-66, civil engineer.
    Eng - 21, 166.

Scott, David, 1838-85, Public Works engr in India.
    PICE - 84, 444-446.
    PIME - 1886, 263-264.

Scott, Edward, 1830-72, engineering manufacturer.
    PIME - 1873, 21-22.

Scott, Edward Baliol,   -1875, Public Works engr in India &
    Ceylon.
    PICE - 46, 276-277.

Scott, Frank Walter, 1864-97, gas engineer.
    PICE - 130, 320.
    PIME - 1897, 142.

Scott, Gideon, 17(82)-1867, contractor & mechanical engr.
    Mech Mag - 86, 263.

Scott, Henry Bowes,   -1894, sanitary & mechanical engineer.
    Eng - 77, 67。

Scott, Maj.-Gen. Henry Young Darracott, F.R.S., 1822-82, Royal
    Engineer(B,D).
    Iron - 21, 383.
    PICE - 75, 319-322.

Scott, James, 18(40)-98, foundry manager。
    TIESS - 41, 382.

Scott, James Portious,   -1892, pig-iron manufacturer.
    Iron - 39, 449*.

Scott, John,   -1874, railway works foreman.
    Eng - 37, 43*.
    Engng - 17, 58*.

Scott, Michael, 1818-90, civil engineer(B).
    JISI - 39, 241-242.
    PICE - 104, 296-301.

Scott, Capt. R   G   ,   -1878, Royal Engineer.
    Eng - 45, 212*.

Scott, William Booth, 1822-91, municipal surveyor(B).
    Iron - 31, 409.
    PICE - 109, 409-411.

Scott-Smith, Capt. H   , 1846-97, cable ship captain.
    Elec - 39, 628-629.

Scriven, William Harry, 1850-98, railway engr in India.
    PICE - 132, 397-398.

Scudamore, Frank Ives, 1823-84, postal engineer(B).
    JIEE - 13, 74.

Seaton, George, 1836-82, water works engineer.
    PICE - 70, 427-428.

Seaward, John, 1786-1858, marine engineering manufacturer(B,D).
    CEAJ - 21, 153.
    PICE - 18, 199-202.

Seaward, Samuel, F.R.S., 1800-42, mechanical & marine engr(D).
    CEAJ - 5, 209.
    PICE - 2, 11-12.

Seddon, John, 1816-91, carriage & wagon manufacturer.
    PIME - 1891, 476.

Seddon, John Frederick, 1848-83, coal owner & mining engr.
    JISI - 23, 668.
    PIME - 1884, 68-69.

Seddon, Robert Barlow, 1853-95, mechanical engineer.
    PIME - 1896, 99.

Seebohm, Henry, 1832-95, steel manufacturer(B,D).
    JISI - 48, 353.

Selby, Col. Henry Oliphant, 18(48)-1900, Royal Engineer.
    Eng - 89, 34.

Sellar, Robert, 1821-84, agricultural implement manufacturer(B).
    Eng - 57, 400.
    Engng - 37, 536.
    Iron - 23, 481.

Sells, Charles, 1820-1900, marine engineer.
    Eng - 90, 566+port.
    Engng - 70, 788-790+port.

Sennett, Richard, 1847-91, Inspector of Machinery to the
    Navy.
    Eng - 72, 211.
    Engng - 52, 336.
    PICE - 97, 409-412.
    PIME - 1891, 476-477.
    TINA - 33, 302-303.

Severn, Henry Augustus, 1833-83, mining engineer.
    PICE - 74, 292-293.

Sewell, John, 1804-87, railway engineer.
    PICE - 92, 401-403.

Seymour, George, 1848-96, mining engineer overseas.
    Elec Rev - 40, 81.
    PICE - 127, 384-385.

Sgouta, Luke George, 1854-96, consulting engr in Greece.
    PICE - 126, 406.

Shanks, Andrew, 1805-69, tool manufacturer.
    PIME - 1870, 16-17.

Shapton, William, 1845-96, mechanical engineer.
    Eng - 82, 211.
    Engng - 62, 286, 316.
    PIME - 1896, 260.

Shardlow, Ambrose, 1842-94, manufacturing engineer.
    PIME - 1894, 599.

Sharpe, Peter, 18(34)-92, ironmaster.
    Iron - 40, 578.

Shaw, Frank, 1840-67, engineering works manager.
    PIME - 1868, 18.

Shaw, Henry Russell,   -1887, rly director in S.America.
    PICE - 90, 443-444.

Shaw, James, 1836-83, ironmaster & contractor(B).
    Engng - 35, 510*.
    Iron - 21, 458.
    JISI - 23, 665.
    PICE - 72, 320-323.

Shaw, John, 1839-87, railway company manager(B).
    PICE - 90, 455-457.

Shaw, William, 1826-96, steel manufacturer.
    JISI - 50, 261*.

Sheldon, W   , 1803-83, omnibus engineer(B).
    Iron - 22, 552.

Shelley, Charles Percy Bysshe, 1827-91, Prof. of Manufacturing
    Art & Machinery, King's College, London(B).
    Eng - 72, 12.
    Engng - 52, 20.
    PICE - 106, 341-343.
    PIME - 1891, 292-293.

Shenton, James, 1839-98, boilermaker.
    JISI - 54, 330.
    PIME - 1898, 706-707.
    TMAE - 8, 221*.

Sheppard, Frederick Augustus, 1819-84, rly surveyor overseas.
    PICE - 78, 443-445.

Sherriff, Alexander Clunes, M.P., 1812-78, ironmaster(B).
    JISI - 12, 291-292*.
    PICE - 52, 284.

Sherriffs, James, 1801-63, company engineer.
    PICE - 24, 543-544.

Shields, William D   , 1870-96, electrical engineer.
    TIESS - 39, 274.

Shipman, Jabez,   -1893, steel & wire manufacturer.
    Iron - 41, 264.

Shopland, James Rew, 1841-96, consulting civil engineer.
    PICE - 129, 391.

Short, John Young, 1844-1900, shipbuilder & ship owner.
    Eng - 89, 130.
    Engng - 69, 170.
    NECIES - 17, 255.

Shotton, John, 1865-98, railway engineer overseas.
    TSE - 1899, 259.

Shuttleworth, Joseph, 1819-83, agricultural implement manufr(B).
    Eng - 55, 91.
    Engng - 35, 115-116.
    Iron - 21, 119.

Sibley, Charles Knowlsley, 18(20)-49, railway engineer.
    PICE - 9, 106-107.

Sibley, George, 1824-91, railway engineer in India(B).
    Iron - 31, 387.
    PICE - 108, 409-411.

Sibley, Robert, 1789-1849, civil engineer & surveyor.
    PICE - 9, 101-102.

Siddell, Thomas, 1806-94, tinworks manager.
JISI - 46, 265*.

Siebe, Augustus, 1788-1872, submarine engineer(B).
Eng - 32, 289.
Engng - 13, 288.
PICE - 36, 301-302.

Siemens, Sir Charles William, F.R.S., 1823-83, metallurgist
& electrician(B,D).
Elec - 12, 37-38.
Eng - 56, 405-406, 426, 468; 58, 405, 415.
Engng - 36, 477-480, 551.
JIEE - 12, 488; 13,442-462. By J.Munro.
JISI - 23, 651-658.
PICE - 77, 352-372.
PIME - 1884, 69-71.
Prac Mag - 5, 289-291.
TINA - 25, 311-312.

Siemens, Dr Ernest Werner von, 1816-92, electrician.
Elec - 30, 159, 181, 614-615+port.
Elec Rev - 31, 697-699+port.
Eng - 74, 507.
Engng - 54, 727-728, 757-758.
Iron - 40, 512.
PICE - 112, 339-344.
PIME - 1892, 570-571.

Silber, Albert Marcius, 1833-86, inventor(B).
Iron - 29, 451-452.
PICE - 89, 496-498.

Silk, Robert,   -1896, electrical works engineer.
Elec Rev - 39, 671.

Simeon, Lionel Barrington, 1852-96, Public Works engineer in
India.
PICE - 127, 385-386.

Simmons, C   J   ,   -1889, electrical instrument maker.
Elec - 23, 25*.

Simms, David, 1845-96, civil engineer.
PICE - 125, 413.

Simms, Frederick Walter, 1803-64, consulting engr in India(B).
PICE - 25, 519/522.

Simms, William, 1793-1860, instrument maker(B).
PICE - 20, 167-168.

Simon, Henry, 1835-99, inventor.
Eng - 88, 91.
Engng - 68, 149-150.

JISI - 56, 295-296.
PICE - 138, 494-497.
PIME - 1899, 270-272.

Simpkin, Robert, 1811-74, railway engineer.
Eng - 39, 86.

Simpson, David Lees, 1839-88, engineer in New Zealand.
PICE - 94, 317-318.

Simpson, Edward Taylor, 1843-78, railway engr in Brazil.
PICE - 59, 317.

Simpson, James, 1799-1869, hydraulic engineer(B).
PICE - 30, 457-460.

Simpson, James, 1829-89, pump manufacturer.
Eng - 67, 415.
Engng - 47, 558*.
PICE - 98, 400.
PIME - 1889, 749-750.

Simpson, Matthew Henry, 1853-93, metallurgist.
JISI - 43, 173-174₀
PICE - 111, 402-403.

Simpson, Robert, 1818-99, marine engineer.
Eng - 87, 58.

Simpson, William, 1809-64, hydraulic engineer.
PICE - 24, 539-540.
PIME - 1865, 16-17.

Simpson, William Storry,   -1896, municipal surveyor.
PIMCE - 23, 480*.

Sinclair, Robert, 1817-98, railway engineer(B).
Eng - 86, 471.
Engng - 66, 623, 650, 723.
PICE - 135, 353-356.
PIME - 1898, 707-709.

Sinclair, Robert Cooper, 1825-93, consulting engineer.
PIME - 1893, 388-389.

Skinner, Herbert Wilfred, 1866-96, harbour engineer.
PICE - 128, 362.

Slagg, Charles, 1823-93, sanitary engineer(B).
PICE - 113, 356.

Slate, Archibald, 1815-60, mechanical engineer.
PICE - 20, 168-169.

Slate, John, 1849-81, sugar refiner.
    PICE - 67, 408-409.

Slater, Isaac, 1819-85, wagon works manager.
    PIME - 1885, 161-162.

Slaughter, Edward, 1814-91, locomotive manufacturer.
    PICE - 105, 316-317.
    PIME - 1891, 293-294.

Slaughter, Frederick William, 1863-99, marine engineer.
    PICE - 140, 284.

Smallman, Richard, 1816-72, railway engineer.
    PICE - 36, 278-280.

Smart, Mortimer Knight, 1827-57, railway contractor.
    PICE - 17, 104-105.

Smart, William George, 1824-83, railway engineer overseas.
    PICE - 76, 369.

Smellie, Hugh, 1840-91, locomotive engineer.
    Engng - 51, 491.
    PICE - 97, 412-414.
    TIESS - 34, 325.

Smethurst, John, 18(48)-1900, colliery manager.
    JISI - 57, 256-257.

Smiles, Frederick Henry, 1861-95, contractor.
    PICE - 122, 400-402.

Smith, Charles, 1843-82, ship works manager.
    Engng - 34, 120.
    JISI - 21, 654.
    PIME - 1883, 25-27.

Smith, Charles Frederick Stuart, 1828-64, civil & mining engr.
    PICE - 28, 621-622.

Smith, Charles Henry Graham, 1851-84, municipal engineer.
    PICE - 78, 445-446.
    PIMCE - 11, 259-261.

Smith, David,   -1888, shipbuilder.
    Engng - 45, 141.

Smith, Edmund James, 18(16)-80, surveyor.
    PICE - 61, 303-304.

Smith, Edward Fisher,   -1892, colliery agent.
    Iron - 40, 39.

Smith, Edward Pease, 1816-77, civil engineer.
    PICE - 52, 285-287.

Smith, Sir Francis Pettit, 1808-74, inventor of screw pro-
    peller (B,D).
    Eng - 37, 139+port; 84, 297.
    PICE - 40, 261-262.
    Prac Mag - 7, 193-195.

Smith, Geoffrey,   -1899, shipowner.
    TIESS - 43, 366.

Smith, George, 17(82)-1869, Belfast harbour engineer.
    Eng - 28, 380.
    Mech Mag - 91, 444.

Smith, Hamilton, 1840-1900, mining engineer.
    Eng - 90, 65.
    PICE - 142, 380-382.

Smith, Henry, 18(23)-97, iron founder.
    Eng - 84, 356.

Smith, Isaac, 1822-68, mechanical engineer.
    PIME - 1869, 17.

Smith, James, 1789-1850, mechanical & agricultural engineer.
    PICE - 10, 91-94.

Smith, James William, 1839-1900, municipal engineer in Bombay.
    PICE - 143, 321-322.
    TSE - 1900, 268.

Smith, John,   -1874, railway contractor.
    Eng - 38, 74*.

Smith, Sir John, 1822-97, manufacturing engineer(B).
    PIME - 1898, 319-320.

Smith, John, 1829-99, manufacturer.
    PIME - 1899, 619.

Smith, John Chaloner, 1827-95, railway engineer(B,D).
    Eng - 79, 245, 339+port.
    PICE - 122, 386-387.

Smith, Gen. Sir John Mark Frederick, F.R.S., 1790-1874, Royal
    Engineer(B,D).
    PICE - 39, 298-299.

Smith, John Paterson,   -1878, consulting engineer.
    Engng - 26, 296.

Smith, John Pigott, 1798-1861, municipal engineer.
    PICE - 21, 594-595.

Smith, Col. John Thomas, F.R.S., 1805-82, Royal Engr(B,D).
    Iron - 19, 473.
    PICE - 71, 422-424.

Smith, Oliver, 1833-93, telegraph engineer.
    Elec - 31, 367.
    Elec Rev - 33, 156.

Smith, Maj.-Gen. Percy Guillemard Llewellyn, 1838-93, Royal
    Engineer.
    PICE - 113, 378-379.

Smith, Richard, 1783-1868, industrial agent to Earl of Dudley.
    Eng - 26, 73, 89.

Smith, Richard, 18(30)-91, metallurgist.
    Eng - 72, 142.
    JISI - 39, 230-231.

Smith, Dr Robert Angus, F.R.S., 1817-84, alkali inspector(B,D).
    Eng - 57, 373, 432.
    Engng - 37, 450-452.
    Iron - 23, 435.
    TIESS - 27, 215.

Smith, Samuel Joseph, 1829-93, engineer with Local Government
    Board.
    PIMCE - 20, 379-380.

Smith, Sydney, 1803-82, inventor of steam pressure gauge(B).
    Eng - 53, 231.
    Engng - 33, 359.
    Iron - 19, 266.

Smith, Thomas,   -1896, iron founder & manufacturer.
    Eng - 82, 456*.

Smith, Thomas Hitchins,   -1868, railway engr in India.
    PICE - 31, 253-254.

Smith, Thomas Macdougall, 1826-86, mining engineer.
    PICE - 84, 446-449.

Smith, William, 1819-81, machinery manufacturer.
    PIME - 1882, 12.

Smith, William,   -1883, steel works manager.
    JISI - 23, 666.

Smith, William, 1823-92, Public Works engineer in India.
    PICE - 110, 382-383.

Smith, William, 1825-78, engng journalist; consulting engr.
    PICE - 56, 289-290.
    PIME - 1879, 16-17.

Smith, William Warren, 1869-1900, mechanical & mining
    engineer.
    PICE - 142, 390.

Smith, Willoughby, 1828-91, telegraph engineer & cable man-
    ufacturer(B).
    Elec - 27, 320*.

Smyth, Sir Warrington Wilkinson, F.R.S., 1817-90, mining
    engineer(B,D).
    Engng - 49, 761.
    Iron - 35, 558.

Smythies, John Palmer, 1843-94, rly engineer in Argentina.
    PICE - 117, 393-394.

Snell, William Henry, -1890, editor of "Electrician".
    Elec - 24, 473-474.
    Eng - 69, 225*; 70, 18*.

Snow, William, -1897, company secretary.
    Eng - 83, 203*.

Soames, Peter, 1830-76, mechanical engineer(B).
    PICE - 45, 260-261.

Sokell, John Henry, 1846-83, company representative.
    PIME - 1884, 71.

Soldenhoff, Richard de, 1844-94, metallurgist.
    JISI - 46, 265.

Somerville, John, 1832-92, gas works manager.
    PICE - 112, 378-379.

Sopwith, Thomas, F.R.S., 1803-79, civil & mining engineer(B,D).
    Eng - 47, 85.
    Engng - 27, 75-76.
    PICE - 58, 345-358.

Sopwith, Thomas, 1838-98, mining engineer overseas.
    Elec Rev - 43, 249.
    Eng - 86, 133*.
    PICE - 134, 408-412.

Sothert, John Lum, 1829-91, lifting machinery manufacturer.
    PICE - 104, 301-302.

Southern, George William, -1874, Inspector of Mines.
    Engng - 18, 273.

Spark, Henry King, 18(27)-99, coal & ironmaster.
    Eng - 88, 556.

Sparkes, Charles Henry, 1846-93, civil engr & architect.
    PICE - 117, 394-395.

Sparks, Col. John Barnes, 1841-93, Bengal Engineer.
    PICE - 115, 404.

Sparrow, John William, 18(20)-91, forge owner.
    Iron - 37, 513.

Sparrow, William Mander, 1812-81, ironmaster.
    JISI - 19, 576-577.

Speck, Thomas Samuel, 1836-83, locomotive superintendent.
    Engng - 36, 432.
    PICE - 76, 370-371.
    PIME - 1884, 71-72.

Spence, James, 1816-97, admiralty engineer.
    Engng - 63, 175.

Spence, John Charters, 1868-98, mechanical engineer.
    PIME - 1898, 320.

Spencer, Charles Innes,   -1875, rly engr in India.
    PICE - 43, 304-306.

Spencer, Charles Thomas, 1856-88, rly engr in India.
    PICE - 95, 391.

Spencer, Eli, 1823-87, engineering manufacturer.
    JISI - 31, 210-211.

Spencer, George, 1810-89, spring manufacturer.
    PICE - 97, 424-425.
    TSE - 1889, 220.

Spencer, Thomas, 1820-83, iron works & colliery manager.
    JISI - 45, 561.

Spice, Robert Paulson, 1814-89, gas engineer.
    PICE - 97, 413-415.
    PIME - 1889, 340-341.
    TSE - 1889, 220-221.

Spiller, Joel, 1790-1873, marine engineer(B).
    PICE - 38, 296-306.

Spittle, Thomas, 1806-81, iron founder.
    Iron - 18, 443.
    PIME - 1882, 12-13.

Spon, Ernest, 1849-90, explosives expert.
    PICE - 104, 315-316.
    TSE - 1890, 214-215.

Spooner, Charles Easton, 1818-89, rly manager & engr.
    Eng - 68, 451.
    Engng - 48, 638.
    Iron - 34, 467.

Spriggs, Christopher, 1828-77, engineering lecturer.
    PIME - 1878, 15.

Squire, Edward,   -1898, ironmaster.
    JISI - 53, 324*.

Stableford, William, 1815-87, carriage & wagon manufr.
    PIME - 1887, 277-278.

Stafford, William, 1863-99, locomotive works engineer.
    Eng - 88, 484*.
    NECIES - 16, 327.

Standfield, John, 1838-90, contracting civil engineer.
    Eng - 69, 206*.
    Engng - 49, 307-308.
    Iron - 35, 211.
    PICE - 101, 301-303.
    TSE - 1890, 211-212.

Stanford, Walter Halsted Cortis, 1840-89, sanitary engr.
    PICE - 97, 415-417.

Stanley, Douglas Austhwaite, 1838-96, consulting rly engr.
    PICE - 125, 414.

Stannard Harry Laurie, 1859-95, rly engineer overseas.
    Eng - 79, 185.
    Engng - 59, 279.
    PICE - 121, 338-340.

Stannard, Robert,   -1891, railway engineer.
    Iron - 38, 319.

Stanton, Maj.-Gen. Frederick Smith, 1832-92, Royal Engineer.
    PICE - 108, 413-415.

Stark, Alexander, 1851-84, ironworks engineer.
    PICE - 76, 371.

Statham, Thomas Henry, 1810-56, railway engineer.
    PICE - 16, 142-143.

Statter, James Samuel, 1854-80, railway engr overseas.
    PICE - 65, 367-369.

Steel, George,   -1899, ship owner.
    Eng - 88, 397.

Steel, Thomas Dyne, 1822-98, consulting civil & mining engr.
　　Eng - 85, 614.
　　PICE - 133, 406-408.

Steele, Robert, 1791-1879, shipbuilder & marine engineer.
　　Engng - 28, 96-97.

Steell, John, 1828-86, rly engineer in India.
　　PICE - 89, 486-488.

Stenhouse, Joseph, 1833-89, shipbuilder.
　　Engng - 47, 590*.

Stent, Sydney, 1845-98, civil engr & architect in S.Africa.
　　PICE - 133, 409-410.

Stent, William Kitson,　　-1900, civil engineer.
　　Eng - 89, 487*.

Stephen, Alexander, 17(94)-1875, shipbuilder.
　　Eng - 39, 294.

Stephen, Alexander, 1832-99, shipbuilder.
　　Eng - 87, 525.
　　Engng - 67, 685.

Stephen, John, 18(35)-77, electrician.
　　Engng - 23, 467.

Stephen, Lt. V　　, 18(52)-92, naval engineer.
　　Eng - 74, 239.

Stephens, Edward Loney, 18(16)-80, municipal surveyor.
　　PIMCE - 7, 147.

Stephens, Frederick Cook, 1829-89, iron & steel works engr.
　　PICE - 96, 344-345.

Stephenson, George, 1781-1848, railway pioneer(D).
　　CEAJ - 11, 279-300, 329-333+illus., 361-364; 12, 6-7,
　　　　68-72, 103-107, 170-173, 205-209.
　　TNEIMME - 8, 33-81 (with Robert Stephenson).By N.Wood.

Stephenson, Henry Palfrey, 1826-90, gas engineer.
　　PICE - 101, 303-305.
　　TSE - 1890, 212-214.

Stephenson, Hugh,　　-1877, coal owner.
　　Engng - 24, 172*.

Stephenson, Robert, 1803-59, pioneer railway engineer(B,D).
　　CEAJ - 13, 271.
　　Eng - 8, 273, 309, 331, 379; 10, 107.
　　PICE - 19, 176-182.
　　TNEIMME - 8, 33-81 (with George Stephenson).By N.Wood.

Stephenson, Sir Rowland Macdonald, 1808-95, rly engr in India(B).
PICE - 123, 451-462.

Steven, Alexander, 18(21)-91, hydraulic machinery manufr.
Engng - 52, 658.
TIESS - 35, 309.

Stevens, Frederick William , 1848-1900, engineer in India.
Eng - 89, 336.
PICE - 141, 353-354.

Stevenson, Alan, 1807-65, lighthouse engineer(B,D).
PICE - 26, 575-577.

Stevenson, David, 1815-86, lighthouse engineer(B,D).
Eng - 62, 76.
Engng - 42, 96.
Iron - 28, 88.
PICE - 87, 440-443.

Stevenson, George Ernest, 1848-99, gas engineer.
Eng - 88, 447*.
PICE - 139, 368-369.

Stevenson, George Wilson, 1825-89, consulting sanitary engr.
PICE - 99, 364-365.
PIME - 1889, 750-751.

Stevenson, Graham, 18(32)-99, inventor.
Eng - 88, 331.
Engng - 68, 394.

Stevenson, Robert, 1772-1850, lighthouse engineer.
PICE - 10, 94-95.

Stevenson, Thomas, 1818-87, lighthouse engineer(B,D).
Eng - 63, 385.
Engng - 43, 454.
Iron - 29, 406.
PICE - 91, 424-426.

Stewart, Alan, 18(59)-94, municipal engineer.
PIMCE - 20, 383*.

Stewart, Alexander, 1866-1900, municipal engineer.
PICE - 142, 390-391.

Stewart, Allan Duncan, 1831-94, consulting civil engineer.
Eng - 78, 409.
PICE - 119, 399-400.

Stewart, Charles Patrick, 1823-82, locomotive manufacturer.
Eng - 54, 26.
Engng - 34, 44*.

```
        Iron - 20, 30.
        JISI - 21, 653-654.
        PIME - 1883, 27-28.

Stewart, D   Y   , 1813-82, engineering manufacturer.
        Engng - 34,339-340.

Stewart, James, 18(39)-90, tube manufacturer.
        Engng - 49, 60, 118.
        Iron - 35, 100.
        JISI - 36, 178-179.

Stewart, James, Jun., 1863-93, rly engr in South America.
        PICE - 117, 395-396.

Stewart, John, 1811-98, shipbuilder.
        PIME - 1898, 709-710.

Stewart, Peter, 1834-1900, works manager.
        JISI - 59, 320.
        TIESS - 43, 364.

Stiff, William Charles, 1836-96, engineering manufacturer.
        PIME - 1896, 601-602.

Stileman, Francis Croughton, 1824-89, railway engineer.
        PICE - 98, 401-402.

Stirling, James, 1800-76, foundry manager & inventor(B).
        Eng - 41, 42.
        Engng - 21, 29.
        PICE - 44, 221-224.

Stirling, John, 1811-82, railway chairman(B).
        Engng - 34, 141.
        Iron - 20, 105.

Stirling, Patrick, 1820-95, locomotive superintendent(B).
        Eng - 80, 485. Port. 74, 515.
        Engng - 60, 612.
        PICE - 124, 421-424.
        PIME - 1895, 542-543.
        TIESS - 39, 272.

Stirling, Revd Robert, 1790-1878, inventor(B,D).
        Engng - 25, 521.
        JISI - 13, 608-609.

Stobart, Henry Smith,   -1880, colliery owner.
        JISI - 17, 690.

Stockwell, Charles Edward, 1870-99, draughtsman.
        PIME - 1899, 619-620.
```

Stoker, Frederick William, 1848-98, iron & steel works manager.
    Eng - 85, 172.
    JISI - 53, 324-325.
    PIME - 1898, 320-321.

Stokes, Charles Lingard, 1826-64, railway engineer.
    PIME - 1865, 17.

Stokes, James Folliott, 1832-99, rly contractor overseas.
    PICE - 136, 361-362.

Stollmeyer, Andre Vlasini, 1858-91, Public Works engr in Trinidad.
    PICE - 107, 419-420.

Stone, Charles, 1820-1900, railway engr in India.
    PICE - 141, 350-351.

Stone, Thomas William, 1848-91, water engineer in Australia.
    PICE - 109, 424-425.

Stoney, Edward Duncan, 1868-98, civil engineer.
    Eng - 85, 163.
    PICE - 132, 398-399.

Stoney, Francis Goold Morony, 1837-97, sluice engineer.
    Eng - 84, 207+port.
    PICE - 130, 316-318.

Storey, John Henry, 1830-83, brass & copper founder.
    PIME - 1884, 72-73.

Storey, Thomas, 1789-1859, chief railway engineer.
    PICE - 19, 182-183.

Storey, Sir Thomas, 18(24)-98, manufacturer.
    JISI - 54, 330-331.

Storrs, Hubert Townsend, 1872-1900, design engineer.
    PICE - 142, 391.

Stotherd, Maj.-Gen. Richard Hugh, 1828-95, Royal Engr; Director-General, Ordnance Survey(B).
    Elec - 35, 73.
    Elec Rev - 36, 622.
    PICE - 121, 343-344.

Stothert, James Lum, 1829-91, manufacturing engineer.
    PICE - 104, 301-302.

Stourton, Athelstan Philip Joseph, 1868-96, mechanical engr.
    PICE - 127, 393-394.

Strachan, James, 1860-1900, consulting mechanical engr.
PIME - 1900, 630.

Strachan, John Henry, 1864-95, municipal engineer.
PICE - 121, 340.
PIMCE - 21, 328-329.

Strapp, Charles Leopol, 1867-95, rly engr overseas.
PICE - 124, 434.

Street, George Edmund, 1824-81, architect.
Eng - 52, 461.

Strick, Richard J   , 1850-97, colliery engineer.
TFIME - 14, 173.

Stringer, William, 1852-93, mechanical engineer.
PIME - 1893, 498.

Strong, Joseph Frank, 1826-95, railway engr in India.
PICE - 120, 365-366.

Stroudley, William, 1833-89, locomotive superintendent(B).
Elec Rev - 25, 721.
Eng - 68, 539.
Engng - 48, 745-746.
Iron - 34, 547.
JISI - 36, 182-184.
PICE - 99, 365-372.
PIME - 1889, 751-752.
TINA - 31, 289.
TSE - 1889, 221-222.

Struvé, William Price, 1809-78, consulting mining engineer(B).
PICE - 52, 278-280.

Strype, William George, 1847-98, mechanical engineer.
PICE - 134, 412-413.
PIME - 1898, 321-322.

Stuart, John Percy, 1861-95, contractor.
PICE - 122, 402.

Stuart, William, 1773-1854, contractor(B).
PICE - 14, 136-137.

Stuart, Maj. William Swainson, 1814-82, Royal Engineer.
PICE - 71, 424-425.

Stubbs, Thomas, 1836-70, locomotive works manager.
PIME - 1871, 18.

Stubbs, William Henry, 1847-90, railway engineer.
PICE - 102, 331.

Sturrock, John,   -1889, Lloyd's surveyor.
     Engng - 47, 6*.

Sulivan, Arthur, 1852-85, Public Works engr in India.
     PICE - 82, 387-388.

Summers, Thomas, 1825-89, marine engineering manufacturer.
     Eng - 67, 352-353.
     PICE - 96, 345-348.

Summers, Thomas, 1855-89, machinery manufacturer.
     PICE - 99, 379.

Summerson, Thomas, 1810-98, gas works engr in Brazil.
     Eng - 86, 587.

Sumner, George Harlowe, 1853-95, mechanical engr in Brazil.
     PICE - 124, 443-444.

Sutcliffe, Alfred,   -1896, municipal surveyor.
     PIMCE - 23, 480*.

Sutherland, 3rd Duke of, F.R.S., 1829-92, industrialist(B,D).
     Eng - 74, 286-287.
     JISI - 42, 299-301.
     PICE - 111, 359-364.
     TIESS - 36, 317*.

Swale, Gerald, 1865-98, electrical engineer.
     PIME - 1898, 710-711.

Swallow, J   , 18(21)-93, ---
     TMAE - 3, 286.

Swallow, James Stuart, 1849-90, railway engr overseas.
     PICE - 101, 306-307.

Swan, John George, 1838-1900, coal & ironmaster.
     Eng - 90, 650.
     JISI - 59, 320-321.

Swansea, 1st Baron, 1821-94, ironmaster(B).
     Eng - 78, 532.

Swanwick, Frederick, 1810-85, railway engineer.
     Eng - 60, 399.
     Iron - 26, 485.
     PICE - 85, 401-407.
     TCMCIE - 15, 33-39.

Swarbrick, Samuel, 1819-99, railway manager(B).
     Eng - 87, 109.
     PICE - 136, 363.

Swetenham, Maj. George, 1856-78, Royal Engineer.
PICE - 55, 334.

Swindell, Charles Evers, 18(19)-91, coal & ironmaster.
Iron - 37, 513.

Swingler, Thomas, 1820-73, ironmaster.
PIME - 1874, 24-25.

Sykes, William, 1815-72, railway engineer.
PICE - 36, 305.

Sylvester, John, 1798-1852, heating & ventilating engr(B).
PICE - 12, 165-167.

Symes,James,Page, 1842-95, engineer & shipbuilder.
PICE - 120, 366-367.

Symington, Hugh, 18(32)-90, railway contractor.
Engng - 49, 676.

Symington, William, 1802-67, inventor.
Mech Mag - 86, 343.

Syms, William, 1820-94, gas works manager.
TSE - 1894, 253.

Synott, Robert Henry Inglis, 1837-72, civil engineer.
PICE - 36, 306.

Tait, James, 1840-99, consulting civil & mining engr.
TIESS - 42, 410-411.

Tait, James, 1843-96, quarry manager.
JISI - 50, 261*.

Talman, James John, 1848-92, Colonial Engr in Lagos.
PICE - 110, 389-390.

Tanner, Henry Thomas, 1842-81, Public Works engr in India.
PICE - 71, 417.

Tanner, Thomas Lanfear, 1844-81, railway engr in India.
PICE - 67, 410-411.

Tannett, Thomas, 1810-77, machine tool manufacturer.
    PIME - 1878, 15-16.

Tansley, Edward, 1832-88, telegraph engineer.
    Elec Rev - 23, 167.

Tarbotton, Mariott Ogle, 1832-87, consulting municipal engr.
    Eng - 63, 218.
    Engng - 43, 229*.
    Iron - 29, 208.
    PICE - 91, 426-429.

Target, John Lewis Felix, 1829-94, Public Works engr in
    Jamaica.
    PICE - 119, 400-401.

Tarrant, Charles, 1815-77, railway engineer.
    PICE - 52, 280-281.

Tasker, Frederick, 1861-97, mechanical engineer.
    PIME - 1897, 517.

Tasker, James Hunter, 1822-51, railway manager.
    PICE - 11, 109-110.

Tasker, John, 18(19)-95, electrical engineer.
    Elec - 34, 388*.
    Elec Rev - 36, 167.
    Eng - 79, 96.

Tatam, Edward John, 1844-98, consulting civil engineer.
    PICE - 133, 410.

Tate, George,   -1866, civil engineer.
    Eng - 21, 166*.

Tate, William, 1838-97, colliery manager.
    TFIME - 16, 126-127.

Taunton, John Hooke, 1821-92, canal engineer.
    PICE - 112, 360-364.

Taylor, Alfred, 18(24)-85, coal owner & manufacturer.
    Iron - 25, 33.

Taylor, Charles Dyke, 1845-76, civil & mining engineer.
    PIME - 1877, 22.

Taylor, Frederick William, 1807-75, engineer in Egypt.
    PICE - 43, 318-319.

Taylor, George, 1820-75, ironmaster.
    PICE - 42, 265.
    PIME - 1876, 24-25.

Taylor, James, 1817-94, government contractor.
    Eng - 78, 232.
    Engng - 58, 389.

Taylor, James, 18(37)-91, ironmaster.
    TMAE - 2, 281.

Taylor, John, 1808-81, mining engineer overseas.
    PICE - 70, 428-430.
    PIME - 1882, 13-14.

Taylor, John, 1817-91, consulting water engineer.
    Iron - 31, 541.
    PICE - 109, 411-413.

Taylor, John, 1836-91, waterworks engineer.
    Engng - 52, 698.

Taylor, Joseph, 1814-90, armaments manufacturer.
    PIME - 1890, 556-557.

Taylor, Norman Maughan, 1868-96, railway engr overseas.
    PICE - 126, 407.

Taylor, Philip, 1786-1870, manufacturer & inventor(B).
    Eng - 30, 20.

Taylor, Richard, 1810-83, mining engineer.
    PIME - 1884, 73-74.

Taylor, Samuel, 1822-71, ironmaster.
    PIME - 1872, 21-22.

Taylor, Thomas Albert Oakes, 1849-94, iron & steel manufr.
    JISI - 47, 263-264*.
    PIME - 1894, 466.

Taylor, Thomas Hardy,   -1869, surveyor.
    PICE - 31, 254.

Taylor, Thomas John, 1810-61, mining engineer.
    TNEIMME - 9, 237-245. By Nicholas Wood.

Taylor, William, 1858-98, draughtsman.
    TIESS - 41, 382.

Terry, Alfred, 1838-79, railway contractor.
    PICE - 56, 290-292.

Thomas, Edwin, 1829-97, canal engineer.
    Eng - 84, 499.
    PICE - 131, 377-379.

Thomas, J E , 18(41)-1900, civil engr & surveyor.
    Engng - 69, 263.

Thomas, James Donnithorne, 1843-1900, mining engineer.
    PICE - 143, 336-337.
    PIME - 1900, 630-631.

Thomas, Hon. John Henry, 1826-84, Director of Public Works,
    Western Australia.
    PICE - 80, 333-335.

Thomas, Joseph,   -1874, ironworks manager.
    Engng - 18, 244*.

Thomas, Sidney Gilchrist, 1850-85, steel manufacturer(B,D).
    Eng - 59, 131-133.
    Engng - 39, 146.
    Iron - 25, 111.
    JISI - 27, 529-533.

Thomas, Thomas, 1833-1900, ironworks & consulting engr.
    PIME - 1900, 631.

Thomas, Walter, 1853-96, municipal engineer.
    PICE - 127, 394-395.
    PIMCE - 23, 480*.

Thomas, William,   -1892, tinplate manufacturer.
    Iron - 39, 449*.

Thomasson, Lucas, 1868-98, mechanical engineer.
    PICE - 135, 369-370.
    PIME - 1898, 542-543.

Thompson, Alfred, 1818-64, consulting civil engineer.
    PICE - 24, 544- 545.

Thompson, David, 18(39)-95, manufacturer.
    Engng - 60, 422.

Thompson, George, 18(04)-95, ship owner.
    Engng - 59, 507.

Thompson, George, 1812-87, forge owner.
    JISI - 31, 211-212.

Thompson, George, 1839-76, engr in South America(B).
    PICE - 45, 261-262.

Thompson, John, 1832-87, ironworks manager.
    JISI - 31, 209-210.

Thompson, John, 1846-1900, municipal surveyor.
    PICE - 141, 354.

Thompson, Sir Matthew William, Bart., 1820-91, railway chairman(B).
    Engng - 52, 653.

Thompson, Robert, 1833-92, municipal surveyor.
    PIMCE - 19, 364-365.

Thompson, William, M.P., 1793-1854, ironmaster(B).
    PICE - 14, 152-155.

Thomson, David, 1816-86, hydraulic engineer.
    PICE - 86, 363-366.
    PIME - 1886, 533-535.

Thomson, George, 1816-95, coal & ironmaster.
    JISI - 47, 264.

Thomson, Grahame Hardie, 18(41)-98, engineering manufr.
    JISI - 54, 331.

Thomson, J   Mann, 18(37)-1900, iron works manager.
    JISI - 55, 267*.

Thomson, James,   -1870, shipbuilder.
    Engng - 9, 177*.

Thomson, Dr James, F.R.S., 1822-92, Prof. of Civil Engng &
    Mechanics, Glasgow University(B,D).
    Elec - 29, 27.
    Elec Rev - 30, 604, 644*.
    Eng - 73, 413, 441.
    Engng - 53, 595-596, 619.
    Iron - 39, 429.
    TIESS - 35, 309-310.

Thomson, John,   -1866, civil engineer.
    Eng - 22, 290.

Thomson, John, 1837-87, gas & water engr; ironfounder.
    Engng - 43, 447, 482.
    JISI - 31, 208-209.
    TIESS - 30, 312-313.

Thomson, John Geale, 1816-60, civil engineer.
    PICE - 20, 158-159.

Thomson, John Turnbull, 1820-84, engr in New Zealand.
    Engng - 38, 589.

Thomson, Michael Nicholson, 1864-95, civil engr & architect.
    PICE - 124, 435.

Thomson, Peter, 1815-76, rly contractor & ironmaster(B).
    PICE - 44, 235-237.

Thomson, Robert S   , 1832-93, engineer in Russia.
    TIESS - 36, 322.

Thomson, Robert William, 1822-73, inventor of rubber tyres(B).
    Eng - 35,156*.
    Engng - 15, 182.

Thorburn, Thomas Charles, 18(24)-1900, municipal engr.
    PIMCE - 26, 252-253.

Thorman, Edward Henry, 1816-91, gas works manager.
    PICE - 109, 425.

Thorn, Peter, 1822-71, contractor.
    Eng - 31, 92*.

Thorneycroft, George Benjamin, 1791-1851, ironmaster(B).
    PICE - 11, 110-112.

Thorneycroft, Thomas, 1816-85, amateur engineer.
    Engng - 40, 260.

Thornhill, John, 1853-96, Public Works engr in India.
    PICE - 129, 396-397.

Thorold, William, 1798-1878, civil engr & surveyor.
    PICE - 55, 321-322.

Thursby, Percy, 1851-98, railway engr overseas.
    PICE - 134, 413.

Thurston, Benjamin Warnes, 1816-85, consulting gas engr.
    PICE - 82, 388-389.

Thwaites, Sir John, 1815-70, Chairman, Metropolitan Board of
    Works(B).
    Eng - 30, 115*.
    Engng - 10, 123.

Thwaites, William Henry, 1850-82, manufacturer.
    PIME - 1883, 28.

Tilfourd, George, 1817-92, ironfounder.
    PIME - 1893, 389.

Till, William Spooner, 1830-91, municipal engineer.
    PICE - 132, 388-390.
    PIMCE - 24, 366-368. Port. frontis v.20.

Tinney, William Upton, 1824-92, water & gas engineer.
    PICE - 111, 403.

Tipping, Henry, 1839-96, consulting mechanical engineer.
    PIME - 1896, 99-100.

Tolmé, Julian Horn, 1836-78, consulting engineer.
    PICE - 55, 319-320.
    PIME - 1879, 17-18.

Tolmie, Andrew Denny,   -1895, sales agent.
    JISI - 48, 353*.

Tomkins, Edward, 1845-76, lecturer in engineering.
    PIME - 1877, 23.

Tomkins, William Graeme, 1817-68, civil engineer.
    PICE - 30, 460-461.

Tomlinson, G   W   ,   -1897, iron founder.
    Eng - 84, 223*.

Tomlinson, Joseph, 1823-94, railway engineer(B).
    Eng - 77, 378, 402 +port.
    Engng - 57, 621-622+port.
    PICE - 117, 388-390.
    PIME - 1894, 163-166.

Tomlison, Charles, 1835-81, water works engineer.
    PICE - 69, 421-422.
    TCMCIE - 11, 16-17.

Tomlison, Henry, 1844-84, waterworks manager.
    Eng - 57, 487.
    PICE - 77, 372-373.

Tone, John Furness, 1822-81, railway engineer.
    PICE - 67, 399-402.

Toozs, Robert William Lyons, 1856-95, rly engr in India.
    PICE - 122, 402-403.

Topham, John, 1833-93, consulting engineer.
    PICE - 114, 397-398.

Topping, John, 1836-86, ironfounder & bridgebuilder.
    Engng - 41, 323*.

Tosh, Dr Edmund George, 1847-99, iron & steel manufacturer.
    Eng - 87, 423*.
    JISI - 57, 257.

Tosh, Robert George, 18(60)-91, ironworks manager.
    JISI - 39, 247.

Tough, J   , 18(51)-99, electrical engineer.
    Elec Rev - 45, 46.

Tovey, Col. Hamilton, 1841-89, Royal Engineer.
    PICE - 98, 408-410.

Townsend, George Barnard, 1814-70, railway agent.
    PICE - 31, 254-255.

Townsend, James,   -1887, telegraph engineer.
    Elec Rev - 20, 504.

Townsend, William, 1838-70, mechanical engineer.
    PIME - 1871, 18.

Townshend, Richard, 1807-88, railway contractor.
    Eng - 65, 283.
    Iron - 31, 296.
    PICE - 93, 492-295.

Trathan, James Jenkin, 1822-80, consulting civil engineer.
    PICE - 65, 378-379.

Trayes, Valentine,   -1900, colliery owner.
    Eng - 89, 368*.

Tremenheere, Maj.-Gen. George Borlase, 1809-96, Bengal Engr(B).
    PICE - 127, 396-397.

Trevithick, Francis, 18(12)-77, railway engineer.
    Eng - 44, 314.

Trevithick, Frederick Henry, 1843-93, railway engineer.
    PICE - 115, 396-397.

Trevithick, Richard, 1771-1833, mechanical engineer(D).
    Eng - 55, 128 (memorial).
    Prac Mag - 7, 289-299(memorial).

Trevor, Maj.-Gen. John Salusbury, 1830-96, Royal Engineer.
    PICE - 126, 413-414.

Trew, James Bradford, 1859-94, mechanical engineer.
    PIME - 1894, 466.

Trewhella, Charles Robert, 1865-93, railway engr in Italy.
    PICE - 113, 357-358.

Trickett, John,   -1888, admiralty dock engineer.
    PICE - 95, 392-393.

Tross, Thomas, 1811-66, civil engineer.
    Eng - 21, 129*.

Trow, Edward,   -1899, labour leader.
    Eng - 87, 164.
    Engng - 67, 225, 326.

Troward, Charles, 1829-73, locomotive superintendent.
    PIME - 1874, 25.

Trubshaw, James, 1777-1853, building contractor(B).
    PICE - 14, 142-146.

Tucker, John Scott, 1814-83, Public Works engr in Barbados.
PICE - 71, 418-420.

Tudhope, Alexander, -1892, colliery owner.
Iron - 40, 163*.

Tuley, William, -1898, municipal engineer.
PIMCE - 24, 368*.

Turley, Thomas, 18(08)-79, furnace owner.
Eng - 47, 148*.

Turnbull, Charles Henry, 1845-88, mechanical engineer.
PIME - 1889, 200.

Turnbull, George, 1809-89, railway engr in India(B).
PICE - 97, 417-420.

Turnbull, John, 1807-88, foundry owner.
Eng - 66, 437.
Engng - 46, 506.
PIME - 1889, 200.

Turnbull, Thomas, 1819-92, shipbuilder & ship owner.
Iron - 39, 386.

Turner, Albert Harrison, 1857-85, gold miner in W.Africa.
PICE - 84, 451.
PIME - 1886, 123.

Turner, Barrow, 1850-88, engineering manufacturer.
PICE - 94, 323-324.

Turner, Frederick Thomas, 1812-77, railway surveyor(B).
PICE - 50, 181-184.

Turner, George Reynolds, 1826-93, railway wagon manufacturer.
PIME - 1893, 389-390.
TIME - 7, 686.

Turner, George William, 1842-96, colliery owner.
TFIME - 12, 126.

Turner, Joshua Alfred Alexander, 1848-95, engr in India.
PIME - 1895, 543.

Turton, John, -1887, Inspector of Mines.
Iron - 30, 35.

Turton, Thomas, 1833-86, forge owner.
JISI - 29, 802-803.
PIME - 1887, 151-152.

Tuxford, William Wedd, 1781-1871, agricultural machinery
    manufacturer(B).
    Eng - 32, 122*.

Tweddell, Ralph Hart, 1843-95, engineering manufacturer.
    Eng - 80, 237.
    Engng - 60, 307, 335-336.
    NECIES - 12, 249-251.
    PICE - 123, 437-440.
    PIME - 1895, 544-546.

Twelvetrees, R H   ,   -1900, railway manager.
    Eng - 89, 119*.

Twibill, Joseph, 1800-85, mechanical engineer.
    Engng - 39, 254*.

Twynam, George Albert, 1843-77, railway engr overseas.
    PICE - 53, 293-294.

Tylden-Wright, Charles Collins Onley, 1832-1900, coal agent(B).
    JISI - 58, 392-393.
    PICE - 142, 382.

Tyler, William James,   -1896, company secretary.
    Elec Rev - 39, 671.

Tyndall, George Reavely, 1856-88, general engineer.
    PICE - 96, 351.

Tyndall, John, F.R.S., 1820-93, Prof. of Natural Philosophy,
    Royal Institution(B,D).
    Elec - 32, 141-142.
    Elec Rev - 33, 610-611, 617.
    Eng - 76, 525.
    Engng - 56, 704-705.
    PICE - 116, 340-351.
    Prac Mag - 7, 353-357.
    TIESS - 37, 195.

Tyzack, William Alexander,   -1889, agricultural implement
    manufacturer.
    Iron - 35, 17.

Udall, Thomas, 18(44)-85, manufacturer.
    JISI - 27, 538-539.

Underhill, George L   ,   -1881, ironmaster.
    Iron - 17, 65*.

Underwood, John, 1841-93, railway construction engineer.
    Eng - 76, 199.

Ure, John Francis, 1820-83, engr on rivers Clyde & Tyne.
    Eng - 55, 366.
    Engng - 35, 448, 480-482.
    Iron - 21, 458.
    PICE - 73, 370-376.

Usher, George Moon, 1828-69, company secretary.
    PIME - 1870, 17*.

Utley, Samuel, 1836-1900, consulting engineer.
    PICE - 142, 382-383.

Vaizey, John Leonard, 1871-96, railway engineer.
    PIME - 1896, 260.

Valentine, Arthur, 1839-72, civil engineer.
    PICE - 36, 306.

Valentine, Charles J   , 18(37)-1900, iron & steel manufr.
    JISI - 58, 393.

Valentine, John Sutherland, 1813-98, civil engineer.
    Engng - 66, 425.
    PICE - 132, 390-392.

Vansittart, John Pennefather, 1837-96, rly engr in India.
    PICE - 89, 488-489.

Van Tromp, B   H   ,   -1899, company director.
    Elec Rev - 45, 606*.

Varley, Cromwell Fleetwood, F.R.S., 1828-83, electrician(B,D).
    Eng - 56, 191.
    Engng - 35, 222.
    Iron - 22, 227.
    JIEE - 12, 456.

PICE - 77, 373-381.

Varley, John, 1828-92, draughtsman.
Iron - 40, 472.
JISI - 40, 298.
PIME - 1892, 410。

Vaughan, Henry, 1829-98, works manager.
Eng - 85, 583.
Engng - 65, 791。

Vaughan, John, 1799-1868, ironmaster(B).
PICE - 28, 622-627.

Vaughan, Thomas, 1834-1900, ironmaster.
Eng - 90, 578.
Engng - 70, 737.
JISI - 58, 393.

Vawdrey, William, 1840-95, waterworks engineer.
Eng - 79, 37.
PICE - 120, 367-368.

Vawser, Robert, 1841-89, municipal engineer.
PICE - 99, 372-373.
PIMCE - 16, 269-270. Port in v.21.

Veitch, Douglas d'Arcy Wilberforce, 1846-83, railway engr.
PICE - 72, 316.

Vereker, Capt. Hon. Foley Charles Prendergast, R.N., 1850-
1900, hydrographer.
Eng - 90, 417*.
PICE - 143, 342-343.

Vernon, Henry, 1814-94, telegraph engineer.
Elec - 32, 480.

Vernon, Herbert Charles Erskine, 1851-93, Public Works engr
in India.
PICE - 114, 388-389.

Vetch, Capt. James, F.R.S., 1789-1869, Royal Engineer; consulting engineer to the Admiralty(B,D).
PICE - 31, 255-262.

Vickers, Edward, 18(04)-97, steel manufacturer.
Eng - 83, 307.

Vickers, George, 18(27)-95, ---
TMAE - 5, 265.

Vickers, George Naylor, 1830-89, metallurgist & entrepreneur.
PICE - 96, 352.

Vidler, Major, 1798-1880, surveyor.
PICE - 61, 299-301.

Vignoles, Charles Blacker, F.R.S., 1793-1875, civil engr(B ,D).
Eng - 40, 359.
Engng - 20, 400-402.
PICE - 43, 306-311.
PIME - 40, 359, 373.

Vignoles, Henry, 1827-99, railway engineer.
Eng - 87, 625.
PICE - 137, 435-437.

Vignoles, Hutton, 1824-89, railway engr overseas.
PICE - 100, 406-408.

Vose, H J , 18(50)-96, telegraph engineer.
Elec Rev - 38, 574.

Vulliamy, Benjamin Lewis, 1780-1854, horologist.
PICE - 14, 155-159.

Waddell, John, 1828-88, contractor.
Engng - 45, 89.
JISI - 32, 222-224.
PIME - 1888, 156-157.

Waddington, Thomas, 1825-69, railway plant manufacturer.
PIME - 1870, 17.

Wade, William Burton, 1832-86, railway engr in Australia.
Eng - 87, 443-446.

Wailes, Edmund Frederick, 18(50)-95, naval architect.
NECIES - 12, 251-252.

Wainwright, William, 1833-95, railway engineer.
PICE - 122, 388.

Wait, James, 1823-99, ship owner.
NECIES - 15, 267.

Wait, John Hooper, 1847-76, engineer in India.
PICE - 45, 262.

Wake, Rowland George, 1854-83, colliery engr in India.
    TCMCIE - 13, 12.

Wakefield, Henry, 1841-99, civil engineer.
    PICE - 137, 437-438.

Wakefield, John, 1812-82, locomotive engr in Ireland.
    PIME - 1883, 28-29.

Wakeford, John, 1833-88, municipal engineer.
    PICE - 95, 393-394.

Walduck,   ,   -1892, iron merchant.
    Eng - 74, 228*.

Wales, Thomas Errington, 1826-86, Inspector of Mines.
    Iron - 27, 461.
    TCMCIE - 15, 39-40.

Walker, Benjamin, 1821-91, hydraulic machinery manufacturer(B).
    Eng - 71, 298.
    Engng - 51, 472.
    Iron - 37, 342.
    JISI - 39, 235-238.
    PICE - 105, 317-320.
    PIME - 1891, 294-296.

Walker, Charles Clement, 18(24)-97, gas works builder.
    JISI - 51, 316.

Walker, Charles Vincent, F.R.S., 1812-82, electrician(B,D).
    Engng - 35, 18.
    Iron - 20, 549.
    JIEE - 12, 1-3.

Walker, J   S   ,   -1892, foundry owner.
    Eng - 74, 456*.

Walker, James, F.R.S., 1781-1862, civil engineer(B).
    Eng - 14, 240, 262.
    PICE - 22, 630-633.
    TINA - 3, xxv-xxvi.

Walker, James Ralph, 1829-61, dock engineer.
    PICE - 21, 567.

Walker, Philip Billingsley, 1865-1900, telegraph engr in
    Australia.
    JIEE - 30, 1249.

Walker, Ralph Teasdale, 1864-93, engineer in Java.
    PIME - 1893, 499.

Walker, Thomas, 18(16)-87, axle manufacturer.
    Eng - 64, 502*.
    Iron - 30, 550.

Walker, Thomas Andrew, 1828-89, railway contractor(B).
    Eng - 68, 450.
    Engng - 48, 635-636; 49, 676*.
    Iron - 34, 467-468.
    PICE - 100, 416-419.

Walker, William, 1830-94, sugar producer.
    Engng - 57, 615.
    PIME - 1894, 166-167.

Walker, William Hugill, 1828-92, steel manufr & merchant.
    PIME - 1892, 410-411.

Wall, J   C   ,   -1897, railway director.
    Engng - 63, 279.

Wallace, David,   -1877, coal & ironmaster.
    JISI - 11, 538-539.
    PICE - 100, 412.

Wallace, James Douglas, 1858-98, railway engr overseas.
    PICE - 137, 442-443.

Waller, James Henry, 1845-85, railway engineer overseas.
    PICE - 83, 443.

Wallis, George Ambrose, 1840-95, waterworks engineer.
    PICE - 124, 424-426.

Walters, Joseph, 1835-75, colliery underviewer.
    TCMCIE - 7, 12-13.

Walton, Rienzi Giesman, 1842-1900, municipal engineer.
    PICE - 143, 324-326.

Ward, Col. David, 1835-88, Bengal Engineer(B).
    PICE - 93, 501.

Ward, John, 1811-58, mechanic.
    PICE - 21, 567-569.

Ward, Richard James, 1817-81, railway engineer.
    PICE - 65, 370-371.

Warden, William Marsden, 1815-90, manufacturer.
    PICE - 103, 392-393.

Wardle, Charles Wetherell, 1821-88, locomotive manufacturer.
    PIME - 1888, 442.

Warham, John Robson, 1820-86, ironfounder.
    PICE - 84, 451-452.
    PIME - 1886, 264.

Waring, Charles,   -1887, rly contractor overseas(B).
    PICE - 92, 410.

Waring, Charles Henry, 1818-87, coal & ironmaster.
    PICE - 91, 446.

Waring, Herbert Francis, 1859-88, civil engr & land agent.
    PICE - 95, 394.

Waring, Thomas, 1825-91, surveyor & sanitary engineer.
    PICE - 104, 303-304.

Warner, William, 1834-1900, foundry owner(B).
    Eng - 89, 361*, 367.

Warren, John Neville, 1818-61, railway engr overseas.
    PICE - 21, 595-597.

Warwick, John, 18(21)-96, railway telegraph engineer(B).
    Elec Rev - 39, 437.
    Eng - 82, 315.

Wasdell, Thomas,   -1897, water engineer.
    Eng - 83, 172.

Washington, Rear-Adm. John, F.R.S., 1801-63, hydrographer(B,D).
    PICE - 23, 513-515.

Wass, Edward Miller, 1829-86, lead miner & smelter.
    TCMCIE - 15, 41-44.

Waterhouse, Thomas, 1821-1900, inventor & manufacturer.
    PIME - 1900-1, 474.

Watkin, William John Laverick, 1839-77, colliery viewer.
    PIME - 1878, 16-17.

Watkins, Francis,   -1847, instrument manufacturer.
    PICE - 7, 15*.

Watkins, William, 1823-1900, telegraph engineer.
    Elec Rev - 46, 316.

Watson, Frederick Howard, 1863-95, civil engineer.
    PICE - 126, 407.

Watson, J   K   ,   -1891, gas company secretary.
    Engng - 52, 596*.

Watson, Joseph Y   ,   -1888, mining engineer.
    Iron - 31, 526.

Watson, Robert, 1811-82, mining engineer.
    PIME - 1883, 29.

Watson, Robert, 1822-91, rly engineer in Australia.
    PICE - 104, 304-305.

Watson, Thomas Colclough, 1822-90, rly engr overseas.
    PICE - 103, 380-381.

Watson, Thomas Parker, 1828-85, rly engr overseas.
    PICE - 80, 335-337.

Watson, William, 1804-83, ship & railway owner.
    PICE - 75, 322-325.

Watson, William, 1804-85, iron & steel works manager.
    JISI - 27, 542.

Watson, William,   -1900, iron merchant.
    Engng - 69, 750.

Watson, Sir William Renny, 1838-1900, manufr & rly director(B).
    Eng - 89, 385.
    Engng - 69, 481.
    PIME - 1900, 334-335.
    TIESS - 43, 364-365.

Watson, William Stephen, 1865-1900, civil engineer.
    PICE - 143, 337-338.

Watt, Alexander, 1823-92, electrical engineer.
    Elec - 28, 317.
    Elec Rev - 30, 101*.

Watts, Edmund Hannay, 1857-94, coal merchant.
    JISI - 46, 266*.

Weallens, William, 1823-62, engineering manufacturer.
    PICE - 22, 633-634.
    PIME - 1863, 13.

Weatherhead, Patrick Lambert, 1844-88, marine engr in Germany.
    PICE - 92, 403-404.
    PIME - 1888, 157-158.

Weaver, Henry, 1825-91, telegraph company director.
    Elec - 31, 574.
    Elec Rev - 33, 345, 616.
    Eng, 1896, 300.

Weaver, William, 1828-68, engineer in New Zealand.
    PICE - 31, 233-236.

Webb, Edward Brainerd, 1820-79, railway engr & surveyor(B).
    PICE - 57, 311-315.

Webb, Frederick Charles, 18(29)-99, telegraph engineer.
    Elec Rev - 45, 110.

Webb, George Winn, 1850-91, gas & tramway engineer.
    PICE - 107, 420-421.

Webb, Maj. Theodosius, 1817-89, Royal Engineer.
    PICE - 100, 419-421.

Webster, John, 1840-75, manufacturer in Brazil.
    PIME - 1876, 25.

Webster, Joseph, -1900, Post Office telegraph engr.
    Elec Rev - 47, 918.

Webster, Thomas, F.R.S., 1810-75, Secretary, Institution of
    Civil Engineers(B,D).
    Eng - 39, 407.
    Engng - 19, 494-495.

Webster, William, 1819-88, contractor(B).
    PICE - 92, 410.

Weeks, Thomas Samuel, 1838-95, municipal engineer.
    PICE - 124, 435-436.

Weems, John, 18(05)-81, ironmaster.
    Engng - 31, 548.

Weems, William, 18(37)-82, ironmaster.
    Engng - 34, 114.

Welch, George William, -1899, electrical engineer.
    Elec Rev - 44, 561.

Weldon, Walter W , F.R.S., 1832-85, chemical engr(B,D).
    Eng - 60, 237.
    Iron - 26, 289.

Wells, Charles, 1836-90, steel sheet manufacturer.
    PIME - 1890, 293.

Wells, Edward John, -1891, civil engineer.
    TSE - 1892, 243-244.

Wells, Lt.-Col. Henry Lake, 1850-98, Royal Engineer(B).
    Elec Rev - 43, 423*.
    JIEE - 28, 682-683.

Wells, Col. John Neave, 1790-1854, Royal Engineer(B).
    PICE - 14, 159-161.

Wells, W   Lewis, 18(36)-92, telegraph engr overseas.
    Elec - 28, 294.

Wells, William Edwin,   -1898, colliery manager.
    JISI - 54, 33.

West, Francis William Isherwood, 1831-60, railway engr.
    PICE - 20, 159.

West, Theodore, 1826-98, draughtsman.
    Eng - 86, 243*.

West, William, 1801-79, civil & mining engineer(B).
    PICE - 59, 308-313.

West, William,   -1893, mining engr & manufacturer.
    Iron - 41, 88.

Western, Col. James Rogers, 1812-71, Bengal Engineer.
    PICE - 33, 273-275.

Westhorp, Thomas,   -1886, ---
    TIESS - 29, 221*.

Weston, William, 1828-86, mining engineer.
    TCMCIE - 16, 14-15.

Westwood, Joseph, 1844-98, engineering manufacturer.
    PICE - 133, 411-412.
    PIME - 1898, 322-323.
    TSE - 1898, 236.

Whalley, Arthur John, 1832-76, railway engineer.
    PIME - 1877, 23-24.

Wheatcroft, Ernest, 18(59)-99, industrial chemist.
    Engng - 67, 749.
    JISI - 55, 267.

Wheatley, Thomas, 1821-83, railway manager.
    Engng - 35, 299-300.
    PIME - 1884, 74.

Wheatstone, Sir Charles, F.R.S., 1802-75, inventor of telegraph;
    Prof. Experimental Philosophy, King's College,London(B,D).
    Eng - 40, 265, 306.
    Engng - 20, 344-345.
    JSTE - 4, 319-334.
    PICE - 47, 283-290.
    Prac Mag - 5, 353-356.

Wheeler, William Herbert, 1867-96, harbour engineer.
    PICE - 126, 408.

While, Charles, 18(17)-95, inventor; manager.
Eng - 79, 474.

Whishaw, Francis, 1804-56, railway engineer(B).
PICE - 16, 143-150.

Whitaker, Charles Woodley, 1837-84, consulting civil engr.
PICE - 79, 369-370.

White, Charles Harcourt,   -1884, railway surveyor.
PICE - 79, 373.

White, George Frederick, 1816-98, cement manufacturer.
PICE - 134, 419-420.

White, James,   -1884, instrument maker.
Engng - 38, 245.

White, John, 1867-98, municipal engineer.
PICE - 136, 362-363.
PIMCE - 25, 478-479.

White, Lawrence Fletcher, 1864-1900, surveyor.
PICE - 144, 318-319.

White, William, 18(10)-81, mining chemist.
Iron - 17, 100.

Whitehead, William, 1829-70, manufacturer.
PIME - 1871, 19.

Whitehouse, Cornelius,   -1883, inventor.
Engng - 35, 152.

Whitehouse, Daniel, 1826-94, tinworks owner.
JISI - 45, 394*.

Whitehouse, Edward Orange Wildman, 18(17)-90, electrical engr.
Elec - 24, 319.

Whitelaw, Alexander, M.P., 1823-79, ironmaster(B).
Engng - 28, 30.
JISI - 14, 327.

Whitfield, John, 1827-84, Public Works engr in India.
PICE - 77, 381-383.

Whitham, Joseph, 1833-66, ironmaster.
PIME - 1867, 16.

Whitley, Joseph, 1816-90, bronze founder.
JISI - 39, 229-230.
PIME - 1891, 193.

Whitmore, John, 1801-72, agricultural engineer.
    Eng - 32, 406*.

Whittaker, John W   ,   -1898, official, Amalgamated
    Society of Engineers.
    Eng - 86, 553.

Whittem, Thomas Sibley, 1835-98, colliery engineer.
    PIME - 1898, 138-139.

Whittington, William, 1840-86, municipal engineer.
    PIMCE - 13, 316.

Whitton, John, 1819-98, railway engr in New South Wales.
    PICE - 132, 393-394.

Whitwell, Thomas, 1837-78, metallurgist(B).
    Eng - 46, 91.
    Engng - 26, 112.
    Iron - 12, 177.
    JISI - 15, 604-608.
    PIME - 1879, 18-19.

Whitworth, Sir Joseph, Bart., F.R.S., 1803-87, mechanical
    engineer & manufacturer(B,D).
    Eng - 63, 75-76.
    Engng - 43, 87-89.
    JISI - 3, 305-307.
    PICE - 91, 429-446+frontis.
    PIME - 1887, 152-156.
    Prac Mag - 3, 322-324+port.
    TIESS - 30, 309-310.
    TSE - 1887, 265-266.

Whyte, Henry Frederick, 1832-83, railway engineer.
    PICE - 75, 308.

Whytehead, William H   Keld, 1825-65, chief engineer to govern-
    ment of Paraguay.
    PIME - 1866, 14.

Wickes, Thomas Haines, 1840-99, Public Works engr in India.
    PICE - 139, 371-372.

Wickham, Henry Wickham, M.P., 1800-67, rly director & iron-
    master(B).
    PIME - 1868, 18*.

Wickham, Lamplugh Wickham,   -1883, ironmaster.
    PIME - 1884, 75.

Wicksteed, John Hamilton, 1851-81, hydraulic engr in S.Africa.
    PICE - 67, 413-415.

Wicksteed, Thomas,1806-71, sanitary & water engineer(B).
Eng - 32, 383.
PICE - 33, 241-246.
PIME - 1872, 22-24.

Wigan, Leonard, 1857-98, contracting municipal engineer.
PICE - 137, 443.

Wight, Theophilus George, 1825-88, surveyor overseas.
PICE - 94, 324-325.

Wightman, Arthur, 1808-64,railway manager.
PICE - 25, 531.

Wightman, William Fawcett, 1844-78, railway engr in Brazil.
PICE - 56, 286-288.

Wigzell, Eustace, 1822-99, mechanical engineer.
Eng - 88, 351*.
Engng - 68, 459.
JISI - 56, 296.

Wild, John, 1862-95, colliery manager.
PIME - 1895, 312.

Wilde, Francis Samuel, 1857-98, railway engr in India.
Eng - 86, 87*.
PICE - 134, 416.

Wildridge, John, 1849-1900, naval architect & consulting engr.
PIME - 1900, 631.

Wilds, W   H   ,   -1890, municipal engineer.
PIMCE - 16, 276*.

Wilkie, George, 1820-57, surveyor & engineer.
PICE - 17, 105-106.

Wilkieson, Maj.-Gen. Charles Vaughan, 1825-78, Royal Engr.
PIME - 1879, 19.

Wilkins, Robert, 1799-1856, mechanical engineer.
PICE - 16, 170.

Wilkinson, George,   -1889, mining engineer.
Engng - 47, 255*.

Wilkinson, John Sheldon, 1837-80, railway engineer.
PICE - 63, 320-321.

Wilks, Clement, 1819-71, engineer in Australia.
PICE - 33, 275-276.

Willans, Jacob G   , 1817-84, inventor & metallurgist.
    JISI - 25, 562-563.

Willans, John William, 1843-95, contractor.
    Elec - 34, 634.
    JISI - 48, 354.

Willans, Peter William, 1851-92, steam engine manufr.
    Elec - 29, 91.
    Elec Rev - 30, 676.
    Eng - 73, 458-459.
    Engng - 53, 655, 661, 686, 693, 720, 747, 789.
    PICE - 111, 395-398.
    PIME - 1892, 224-225.

Willcox, Francis William, 1840-96, marine engr & naval architect.
    NECIES - 12, 253-254.
    PIME - 1896, 100.
    TIESS - 39, 272-273.

Willcox, Joseph,   -1874, ironworks engineer.
    Engng - 17, 283.

Willcox, William, 1830-89, rly contractor & surveyor(B).
    PICE - 100, 408-409.

Willett,John, 1815-91, locomotive engineer(B).
    PICE - 97, 414-416.

Williams,   , 1820-74, locomotive superintendent.
    Eng - 37, 54.

Williams, Alfred, 1826-94, gas engineer(B).
    Eng - 78, 76.
    TSE - 1894, 254-255.

Williams, Charles Wye, 1780-1866, marine engr; ship owner(B).
    Eng - 21, 252*.
    PICE - 28, 627-632.

Williams, Edward, 1826-86, ironmaster.
    Engng - 41, 594.
    Iron - 27, 552.
    JISI - 29, 213-215.
    PICE - 88, 439-442.
    PIME - 1886, 264-265.

Williams, Edward Bentinck, 1847-81, civil engineer.
    PICE - 67, 415.

Williams, Edward Leader, 1802-79, civil engineer.
    PICE - 57, 315-317.

Williams, Israel,   -1894, ironworks manager.
    JISI - 46, 266.

Williams, James, 18(46)-98, colliery owner.
    Eng - 85, 466*.
    JISI - 54, 332.

Williams, John, 1818-87, railway surveyor.
    PICE - 92, 393.

Williams, John,   -1900, foundry manager.
    JISI - 60, 332*.

Williams, John Evelyn, 1845-95, harbour engineer(B).
    PICE - 120, 368-370.

Williams, John Michael, 1813-80, copper smelter(B).
    Iron - 15, 138.

Williams, John Ward, 18(30)-93, foundry owner.
    Iron - 41, 316.

Williams, Joshua,   -1872, tinworks owner & rly manager.
    Engng - 13, 314*.

Williams, Nicholas, 18(40)-99, mining engineer.
    JISI - 55, 268*.

Williams, Owen, 18(18)-93, ---
    TMAE - 3, 286*.

Williams, W   Mattieu,   -1892, metallurgist.
    Iron - 40, 495.

Williams, Walter, 1830-92, coal & ironmaster.
    Iron - 39, 231.
    JISI - 42, 291-293.

Williams, William,   -1891, mechanical engineer.
    Iron - 38, 319.

Williams, William, 1822-96, ironworks engineer.
    Engng - 61, 227.

Williams, William Lawrence, 1848-98, consulting mechanical engr.
    PICE - 135, 358-359.
    PIME - 1898, 711.

Williamson, John Richard Hutchinson, 1859-92, electric light
    manufacturer.
    Elec - 30, 181.
    Elec Rev - 31, 734*.
    PICE - 112, 371-372.

Williamson, Richard,   -1874, municipal surveyor.
    Eng - 37, 295*.

Willis, Arthur,   -1881, industrial chemist.
    JISI - 19, 576.

Willman, Charles, 1832-94, mechanical engineer(B).
    PICE - 118, 464-465.
    PIME - 1894, 279.

Willmott,Arthur Wellesley Westmacott, 1852-86, gun factory
    manager in China.
    PIME - 1886, 265-266.

Willock, Capt. Harry Borlase, 1854-89, Royal Engineer.
    PICE - 96, 352-354.
    PIME - 1889, 200-201.

Willson, John, 18(48)-91, municipal surveyor.
    PIMCE - 18, 441.

Wilson, Engineer-General Alexander, 1776-1866, engr in Russia(B).
    PICE - 30, 461-465.

Wilson, Alexander Hall, 1840-99, shipbuilder.
    TIESS - 43, 365-366.

Wilson, Allan, 1820-97, Public Works engr in India.
    PICE - 131, 379-381.

Wilson, Daniel, 1790-1849, manufacturing chemist.
    PICE - 9, 101-102.

Wilson, Edward, 1820-77, consulting engineer.
    Eng - 44, 164.
    Engng - 24, 212.
    PIME - 1878, 17-18.

Wilson, George,   -1871, railway chairman.
    Eng - 30, 441*.

Wilson, George, 1829-85, iron & steel works chairman.
    Engng - 40, 541.
    Iron - 26, 503.
    JISI - 27, 541-542.
    PICE - 91, 448-449.
    PIME - 1885, 527-528.

Wilson, Guillermo Andres, 1852-93, rly engr in Latin America.
    PICE - 113, 358-359.

Wilson, Isaac, 1822-99, ironmaster & engng manufacturer(B).
    Eng - 88, 318.

Wilson, James, 18(22)-94, mining engineer.
    Eng - 78, 532.
    Engng - 58, 766.

Wilson, James, 1842-1900, water engineer.
    Eng - 90, 86.
    PICE - 143, 326-329.

Wilson, John, 1787-1851, ironmaster.
    PICE - 11, 113-115.

Wilson, John, 1848-98, consulting engineer.
    TIESS - 42, 411.

Wilson, Joseph William, 1829-98, consulting engr & inventor.
    Eng - 86, 524.
    Engng - 66, 679.
    PICE - 135, 359-360.
    PIME - 1898, 712-713.

Wilson, Peter, 18(25)-75, railway engineer.
    Engng - 19, 95.

Wilson, Robert, 1803-82, machine tool manufacturer.
    Eng - 54, 89.
    Engng - 34, 119.
    Iron - 20, 105.
    PIME - 1883, 29-31.

Wilson, Robert, 1851-98, engineering manufacturer.
    PICE - 131, 381-383.
    PIME - 1898, 139-140.

Wilson, Thomas, 1843-90, marine superintendent.
    NECIES - 6, xliii.
    PIME - 1890, 293-294.

Wilson, William, 1822-98, railway contractor(B).
    Eng - 86, 324.
    Engng - 66, 440, 465.
    PICE - 135, 361-362.

Wilson, William John, 1851-1900, water engr in Cairo.
    Eng - 90, 190*.
    PICE - 142, 383-384.

Wimshurst, Henry, 1804-84, shipbuilder.
    Engng - 38, 258.

Winder, Thomas Robert, 1817-83, railway & dock engineer.
    PICE - 74, 290-291.

Wingate, Robert, 1832-1900, railway engineer overseas.
    Eng - 89, 647*.
    PICE - 142, 384-385.

Wingate, Thomas, 1800-69, shipbuilder.
    Eng - 28, 284.

Winsland, Nicholas,   -1846, contractor.
    PICE - 6, 5*.

Winter, George Kift, 1842-98, telegraph engr in India.
    Elec - 40, 507-508.
    Elec Rev - 42, 192.
    Eng - 85, 133.
    Engng - 65, 177.
    JIEE - 28, 683-684.
    PICE - 133, 412-413.

Winteringham, Henry,   -1877, civil engineer in Peru.
    PICE - 50, 184-185.

Wise, Francis, 1820-68, consulting engr & patent agent.
    PIME - 1869, 17-18.

Wise, Thomas,   -1868, engineer & patent agent.
    Engng - 7, 13*.

Wiseman, William, 1843-93, railway engineer in India.
    PICE - 114, 385-387.

Withers, John, 18(14)-57, ---
    PICE - 22, 640*.

Wood, Maj. Alexander,   -1898, telegraph engineer overseas.
    Elec Rev - 41, 152.

Wood, Alfred Hope, 1826-89, gas engineer.
    PICE - 99, 380-381.

Wood, Charles, 18(39)-94, machinery manufacturer.
    Eng - 78, 219.
    Engng - 58, 334.

Wood, Edward Walter Nealor, 1856-89, railway engineer(B).
    PICE - 99, 381-382.
    PIME - 1889, 752-753.

Wood, George, 1826-79, ironmaster.
    JISI - 14, 330.

Wood, Henry, 1805-86, admiralty dockyard engineer.
    PICE - 90, 436-439.

Wood, John,   -1895, colliery owner.
    JISI - 48, 354*.

Wood, Nicholas, F.R.S., 1795-1865, mining engineer(B).
    Eng - 20, 415*; 21, 37.

        Iron - 40, 578.
        PICE - 31, 236-238.
        PIME - 1866, 15.
        TNEIMME - 15, 49-59. By Thomas Doubleday.

Wood, Robert Henry, 1851-95, draughtsman & mechanic.
        PIME - 1895, 313.

Wood, Sancton, 1815-86, railway architect.
        PICE - 86, 376-379.

Wood, Thomas, 1849-86, steel works engineer.
        JISI - 29, 803.
        PIME - 1886, 463-464.

Wood, Thomas Eason,   -1892, telegraph engineer.
        Elec - 28, 343.

Wood, Hon. Walter Abbots, 18(16)-92, agricultural machinery
        manufacturer.
        Iron - 39, 79.

Woodcock, William, 1814-74, heating & ventilating equipment
        manufacturer.
        PICE - 39, 299-300.

Woodcroft, Bennett, F.R.S., 1802-79, inventor; patentee(B,D).
        Eng - 47, 118-119.
        Engng - 27, 140-141.

Woodford, John Wyman, 18(22)-78, inventor.
        Eng - 45, 386.
        Engng - 25, 429.

Woodhead, John Proctor, 1843-84, consulting engineer.
        PIME - 1884, 404-405.

Woodhouse, George, 18(01)-68, civil engineer.
        Eng - 26, 308*.

Woodhouse, John Thomas, 1809-78, consulting mining engineer(B).
        PICE - 55, 321-325.
        TCMCIE - 7, 13-16.

Woodhouse, Otway Edward, 1855-87, electrical contractor.
        Elec - 19, 415*, 435.
        Elec Rev - 21, 324, 347.
        Eng - 64, 296.
        Engng - 44, 398.
        Iron - 30, 354.
        PICE - 91, 454-455.

Woodhouse, Thomas Jackson, 1793-1855, railway engineer(B).
        PICE - 16, 150-154.

Woodhouse, William Henry, 1815-64, telegraph engineer.
    PIME - 1865, 18.

Woodiwiss, Sir Abraham, 1828-84, railway contractor(B).
    Iron - 23, 186.
    TCMCIE - 13, 12-15.

Woods, Henry,   -1899, surveyor.
    PIMCE - 25, 480*.

Woods, Joseph, 1816-49, mechanical engineer.
    PICE - 9, 107-108.

Woolbert, Henry Robert, 1822-75, railway engineer.
    PICE - 42, 265-266.

Woolcock, Henry, 1842-92, civil & mining engineer.
    JISI - 43, 174*.
    PICE - 110, 390-391.

Woolley, Joseph, 1817-89, naval architect.
    Eng - 67,271.
    Iron - 33, 295.
    TINA - 30, 463-465+port.

Wordsworth, Charles Favell Forth, 1803-74, industrial solicitor(B).
    PICE - 39, 300.

Wragge, Frederick, 1821-86, coal & ironmaster(B).
    Engng - 42, 639.
    Iron - 28, 547.
    JISI - 29, 799-800.

Wray, William, 1830-71, shipbuilder.
    PIME - 1872, 24.

Wreathall, Harry, 1866-94, railway engineer.
    PICE - 120, 373.

Wrigg, H   ,   -1879, engr in New Zealand.
    Engng - 28, 122*.

Wright, Alexander, 1816-59, gas engineer(B).
    PICE - 19, 183-185.

Wright, Benjamin Frederick, 1845-88, railway engr in Japan.
    Eng - 65, 277.
    PICE - 94, 318-319.
    PIME - 1888, 265-266.

Wright, E T   ,   -1877, ironmaster.
    Engng - 23, 361*.

Wright, George Hustwait, 1834-89, rly engr overseas.
PICE - 102, 332-334.

Wright, Sir James, 1823-99, admiralty engineer(B).
Eng - 87, 384.
Engng - 67, 517.
TINA - 41, 373.

Wright, Joseph,1826-94, chain & anchor manufacturer.
PIME - 1894, 167-168.

Wright, Owen, 1835-83, manufacturer.
PIME - 1884, 75.

Wright, Peter, 1803-74, manufacturer.
PIME - 1875, 32-33.

Wright, Robert Edwin, 1841-87, Public Works engr in India.
PICE - 94, 320-321.

Wright, Thomas, 1812-91, waterway engineer on Danube.
PICE - 104, 322- 323.

Wrightson, Stephen,   -1898, company secretary.
Eng - 86, 577*.
JISI - 54, 332*.

Wrigley, Francis, 1811-62, consulting engineer.
PIME - 1862, 13.

Wrottesley, Maj. Alfred Edward, 1855-99, Royal Engr; telegraph engineer(B).
Elec Rev - 45, 725.

Wyatt, William John,   -1898, municipal engineer.
PIMCE - 25, 479*.

Wyld, Robert Stodart, 1855-91, water engineer.
PICE - 104, 306-307.

Wyles, Frederick, 18(61)-1900, electrical engineer.
Elec Rev - 46, 955.
Eng - 89, 628*.
JIEE - 29, 957.

Wylie, Alexander, 18(19)-76, shipping engineer.
Engng - 22, 322.

Wylie, George, 1860-96, Public Works engr in India.
PICE - 129, 397.

Wylie, Henry Johnston, 1822-71, railway engineer.
PICE - 36, 280-281.

Wylie, John Condie, 1853-99, mining engineer.
   PIME - 1899, 472.

Wylie, Robert, 1850-86, engineering works manager.
   JISI - 29, 804.
   PIME - 1886, 464.
   TIESS - 30, 313.

Wyllie, Andrew, 1823-1900, manufacturing engineer.
   PIME - 1900, 335-336.

Wynne, Thomas, 1807-91, Inspector of Mines(B).
   PICE - 106, 343-345.

Wythes, George, 1811-83, rly contractor in India(B).
   Engng - 35, 238.
   Iron - 21, 206.
   PICE - 74, 294-297.

Wythes, George Edward, 1841-75, railway director.
   PICE - 41, 232.

Yapp, George Wagstaffe, 1811-80, engineering journalist(B).
   Eng - 50, 414.

Yarrow, Thomas Alfred, 1817-74, consulting engineer.
   PICE - 39, 282-283.

Yates, Henry, 1820-94, railway engr in Canada.
   PIME - 1894, 466-467.

Yeo, Frank Ash, M.P., 1832-88, coal owner(B).
   Engng - 45, 239*.

Yeo, George Jope, 1844-86, gas engineer in Shanghai.
   Eng - 87, 449-450.

Yockney, Samuel Hansard, 1813-93, consulting civil engr(B).
   PICE - 116, 372-374.

Yolland, Col. William, F.R.S., 1810-85, Royal Engr; railway
   inspector(B,D).
   Eng - 60, 206.
   Iron - 26, 244.

York, John Oliver, 1811-87, ironmaster & contractor.
    PICE - 94, 325-326.

Young, Barclay Hughes, 1854-93, Public Works engr in India.
    PICE - 113, 359.

Young, David, 1852-99, patent agent & consulting engineer.
    PIME - 1899, 620.

Young, Francis Mortimer, 1814-60, railway engineer.
    PICE - 20, 159-160.

Young, Dr James, F.R.S., 1811-83, chemical engineer(B,D).
    Eng‾ - 55, 385.
    Engng - 35, 510. see also 5, 544-545.

Young, John, 1826-95, consulting civil engineer & architect.
    PICE - 124, 426-427.

Young, Robert, 1846-99, railway plant contractor.
    JISI - 34, 268.

Younghusband, Lt.-Gen. Charles Wright, F.R.S., 1821-99, arms
    engineer(B).
    Eng - 88, 438.

Younghusband, Oswald, 1833-81, railway engineer overseas.
    PICE - 64, 339-340.

Yule, John, 18(11)-77, engineering manufacturer.
    Engng - 23, 438.

Yule, William, 1823-81, ironfounder in Russia.
    PIME - 1882, 14.

                    ****

                    245

Dübs, Henry.
   add: PIME - 1877, 18-19.

Fairbairn, Sir William.
   add: Prac Mag - 4, 241-245.

Forster, George Baker,   -1900, mining enginer.
   TFIME - 21, 5-6.

Heinke, John William.
   add: PICE - 31, 247-248.

Holst, L   M   .
   add: Elec - 24, 109.

Jackson, Matthew Murray, 1821-93, mechanical engr in Europe.
   PIME - 1893, 91-93.

Marsden, Henry Rowland.
   add: Prac Mag - 6, 1-3.

Merryweather, Henry.
   add: Iron - 19, 16.

Mitchell, Charles.
   add: PIME - 1895, 537-539.

Smith, Willoughby.
   add: Iron - 38, 77*.

Wright, Alfred, 1821-94, civil engineer.
   PICE - 47, 419-421.

*****